FUNGI

DK LYNNE BODDY & ALI ASHBY

FUNGI

DISCOVER THE SCIENCE AND SECRETS
BEHIND THE WORLD OF MUSHROOMS

Marasmius capillaris

CONTENTS

**Amethyst deceiver
(*Laccaria amethystina*)**
This beautifully vibrant mushroom with its deep violet cap and stem can often deceive. As this mushroom dries out it turns a pale grey, almost white – which makes identification difficult.

FUNGAL FILAMENTS

A teaspoon of woodland soil (about 1 g) contains up to 100 m (330 ft) of fungal filaments called hyphae. So, in the soil beneath your feet on a short walk, the total length of hyphae is enough to stretch several times around the Earth's equator.

Fungi were once grouped with plants, but they are now recognized as a separate kingdom of living things. Scientists estimate that there are 5 million species of fungi, ranging from microscopic yeast and networks of fine filaments to mushrooms, puffballs, and brackets.

WHAT ARE FUNGI?

Fungi are neither plants nor animals. They are a separate kingdom of organisms, present in just about every habitat on Earth, from deep sea caves to fruit bowls, and they even form part of the human body's microbiome. Fungi are classed as microorganisms. For example, yeasts form individual cells, mostly 3–4 but up to 40 thousandths of a millimetre (μm) in diameter, whereas filamentous (thread-like) fungi form microscopic tubes called hyphae. These can be made up of one or many cells of varying length, and just a few thousandths of a millimetre thick, around 40 times thinner than a human hair.

When water is added to dried yeast, or when fungal spores are layered on top of one another in a spore print (pp.172–173), they become visible to us; however, we need a microscope to observe them in detail. Hyphae can be seen with the naked eye when they come together en masse, for example, growing on food or on the underside of dead wood. They group together to form visible fruit bodies, and also form networks called mycelium, often microscopic and spreading over large distances. One of the largest living things is the mycelium of a fungus (p.36).

THE MAIN BODY: MYCELIUM The main body of a tree is its roots, trunk, and branches. Likewise, the main structure of a filamentous fungus is its fine hyphae that form networks called mycelia (pp.36–37), which are often hidden from view. Unlike plants, fungi cannot make their own food but instead must find and digest dead organic matter or get it from a living plant or animal (pp.38–39).

FUNGAL FRUITS: MUSHROOMS Apples are the fruits of apple trees, and pears are the fruits of pear trees. The fruits differ structurally, but both produce, protect, and disperse seeds. Similarly, mushrooms, brackets, earth stars, cups, corals, and puffballs (pp.180–213), among many others, are the fruit bodies of certain types of fungi: their role is to produce, protect, and disperse fungal spores, allowing the fungus to spread to new environments.

KEY TERMS
Scientists use specialist words to talk about kingdom fungi, many of which have been included throughout this book. Check the glossary on pages 288–289 for a simple explanation of key terms.

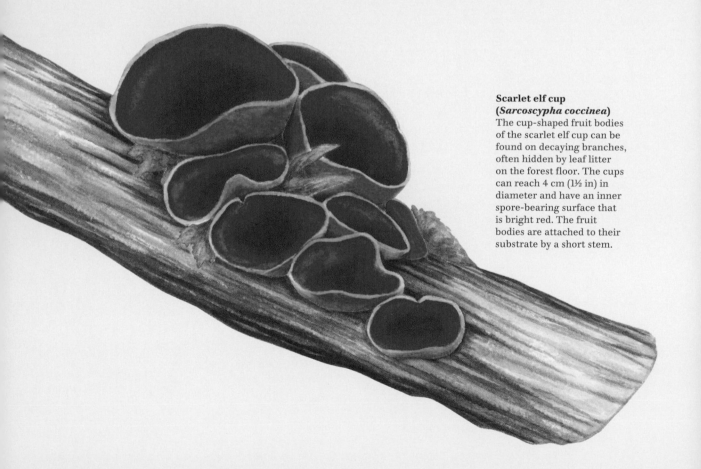

Scarlet elf cup (*Sarcoscypha coccinea*)
The cup-shaped fruit bodies of the scarlet elf cup can be found on decaying branches, often hidden by leaf litter on the forest floor. The cups can reach 4 cm (1½ in) in diameter and have an inner spore-bearing surface that is bright red. The fruit bodies are attached to their substrate by a short stem.

Mycelium of some species produce fruit bodies only once in their life, and some make them for a short time each year for many years, while others make fruit bodies that continue to produce spores for many months or even years.

MORE THAN MUSHROOMS All mushrooms are fungi, but not all fungi are mushrooms. Mushrooms are the fruit bodies of a group of fungi called basidiomycetes (pp.14–15). They form the tip of the proverbial iceberg: it is the mycelium that is the main part of most fungi. Some fungi glow in the dark (pp.124–125), some make music (pp.236–237) or eat roundworms (p.87), and others soak up radiation or break down plastics (pp.280–281), while others produce valuable enzymes used in industry (pp.276–277). Many fungi are eaten by animals (pp.86–87), some are edible to humans (pp.216–217), while others are deadly poisonous (pp.96–97, 178–179). Some mushrooms have been used by humans for centuries for their perceived medicinal properties (pp.248–251), while others have been incorporated into spiritual practices (pp.240–243).

Some fungi can cause disease and even death to plants (pp.72–81) and other organisms (pp.88–91), yet most plants (pp.60–63) and also many animals (pp.82–85) would not survive without the beneficial partnerships they form with fungi. Some fungi break down dead animal and plant material, and even the very fabric of our homes (pp.32–33). They are the Earth's best recyclers: without them, the nutrients in dead trees and other organic matter would be locked up and there would be no fertile soil to grow new plant life (pp.280–281).

This book is not just about mushrooms, and it is not a reference book for identifying mushrooms. It is a book for the inquisitive, for those who want to learn more about "kingdom fungi" - what they are, what they do, and how they have come to shape the world we live in.

Why is this important? Because one thing is certain: we could not survive on Earth without fungi.

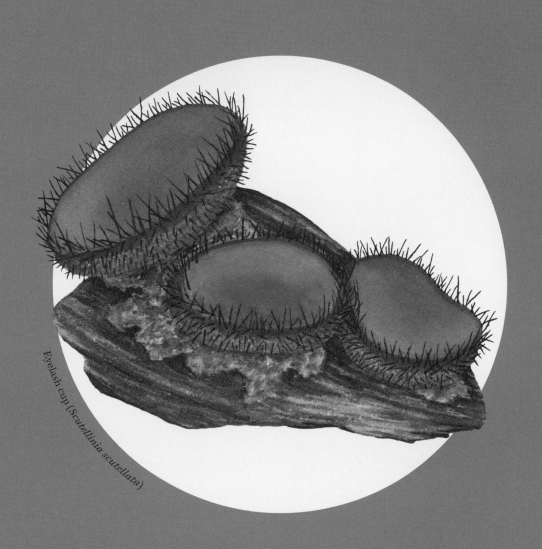

Eyelash cup (*Scutellinia scutellata*)

DISCOVER

CHAPTER I

WHAT IS A FUNGUS? How and when did the fungal kingdom come into being? Understand the important differences between fungi and other organisms, explore how fungi of the world are related, and discover how many species there are and where we can find them.

*Fungi are more closely related to **animals** than they are to plants, and they first made an appearance on Earth about **1 billion years** ago.*

Life, but not as we know it, began on Earth around 3.5 billion years ago with the evolution of simple, single-celled organisms. These evolved into more complex cells, some that would eventually form the basis of fungi.

THE RISE OF KINGDOM FUNGI

The first, simple life forms lived in communities within slimy mats called biofilms. After almost 2 billion years, these single-celled organisms evolved into more complex eukaryotic cells. These cell types have a nucleus (containing genetic material) and compartments called organelles, each equipped for a particular task: the mitochondrion, for example, helps to produce energy. The last eukaryotic common ancestor evolved to have new features that would enable it to transition into the plant, animal, and fungal cell forms we know today.

WHEN DID FUNGI APPEAR? Scientists can estimate when key evolutionary events occurred using the "molecular clock" (the rate at which mutations occur in genetic material) and fossil evidence. These techniques reveal that simple plant forms were the first to split from the common ancestor, followed 600 million years later by fungi and animals. This means that fungi are more closely related to animals than to plants, and they first appeared on Earth around 1 billion years ago.

WHAT WERE THE FIRST FUNGI LIKE? The first fungi were single-celled, water-dwelling organisms, able to form simple structures for bearing spores. Some fungal groups still have features in common with these ancestral forms, including members of the Chytridiomycota (pp.14–15), the Microsporidia, and the Cryptomycota.

PROMISCUOUS LIFE FORMS

In the ancient biofilm communities of early Earth, cells touched, exchanged contents, and even engulfed one another in an evolutionary "trial and error", which resulted in mixing of characteristics. Indeed, some of the most primitive groups, such as the water moulds (Oomycetes, pp.20–21) and the slime moulds (Myxomycetes, pp.20–21), are not considered plant, animal, or fungus.

This kingdom of over 5 million species contains many different types of fungi – from single-celled yeast and microscopic species that grow inside plant or animal cells, to some of the biggest organisms on the planet.

WHO BELONGS TO KINGDOM FUNGI?

CHYTRIDIOMYCOTA

Chytrids are microscopic and live in salt water, freshwater, estuaries, and wet soil. Unlike most fungi, they have a ball-shaped body with hyphae – called rhizoids – that extend from the base. Many grow on dead material, but others grow on or in living plant or animal cells and cause disease.

ZOOPAGOMYCOTA

These fungi specialize in killing invertebrates. They can sometimes be seen when masses of their hyphae protrude from the bodies of insects they have killed, or when showers of spores form a carpet around a dead insect.

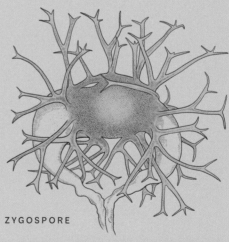

ZYGOSPORE

MUCOROMYCOTA

Mucoromycotina, a group within this phylum, feed on dead plant and animal material. They can be seen when their hyphae grow out and when they produce minute packages of dark spores, such as pin moulds on bread or animal dung (pp.144–145). Glomeromycotina, another group, are important partners with the roots of plants (pp.60–61). They are microscopic and do not produce fruit bodies, but they do make spores in soil.

The fungal kingdom is divided into major groups called phyla (singular: phylum), as are other organisms, such as animals, whose phyla include mammals, birds, reptiles, amphibians, and fish. Each fungal phylum contains closely related species and is subdivided further based on how closely related these species are.

The main fungal phyla are described here. Science still has much to discover about the other types.

ASCOMYCOTA

Ascomycetes are the largest phylum in the fungal kingdom, with 97,000 species described so far. All of them can make prolific numbers of spores without mating. Some can mate and then make spores in a sac-like structure called an ascus (plural: asci), after which the phylum is named. Some fungi produce these in minute fruit bodies, whereas others are visible to the naked eye.

RHODOTUS PALMATUS
WRINKLED PEACH

MICROSTOMA PROTRACTUM
ROSY GOBLET

BASIDIOMYCOTA

So far, 52,000 basidiomycete species have been named and described by science. Most produce large fruit bodies at some point – days, weeks, or even many months – after mating. They form their spores on cells called basidia (singular: basidium), which is where the phylum gets its name. Rusts and smuts (pp.72–73), which are important basidiomycete plant pathogens, do not produce large fruit bodies, but their reproductive structures can nevertheless be seen.

All fungal cells have the same basic structure – whether it is a budding yeast cell, a spore released from a reproductive structure such as a mushroom, or a cell or group of cells that form fungal filaments called hyphae.

THE FUNGAL BODY PLAN

The main structural features of fungal cells include a cell wall made of chitin, which surrounds a cell membrane. Inside are membrane-bound compartments – organelles – which do specific tasks in the cell.

FUNGAL CELLS Fungal cells are eukaryotic, like those of animals, humans, and plants. Eukaryotic cells are more complex than bacterial cells, as they house a membrane-bound nucleus containing genetic material and organelles. However, there are some key differences. Like animal cells, fungal cells do not contain chloroplasts. These organelles, present in plant cells, carry out photosynthesis – a process that uses energy from sunlight to make food. Fungi, like animals, cannot make food themselves, so they must find food and digest it.

HYPHAL CELLS Hyphae grow from the tips outwards, and can form new growth by creating new branches. When they branch, hyphae form a network called mycelium (pp.36–37), the main body of a filamentous fungus. The hyphae of Basidiomycota and Ascomycota (pp.14–15) fungi form cross walls, called septa, at right angles to the growing filament. Each cross wall has a small pore, wide enough for some cell organelles to move between cells, which enables communication between neighbouring cells within the mycelium.

WHAT IS CHITIN?

Chitin is a structural polymer, which is also found in the exoskeletons of insects and crustaceans; this gives an idea of how tough this structural polymer is. In the fungal cell wall, chitin is partly responsible for hyphal strength, enabling the coordinated masses of hyphae to push up and emerge from beneath the forest floor, through the trunk of a tree, and even through tarmac.

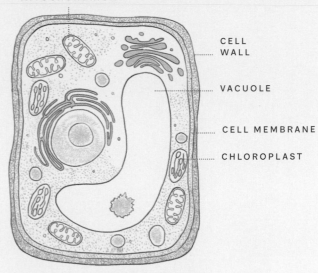

CELL MEMBRANE

MITOCHONDRION

MITOCHONDRION

CELL WALL

VACUOLE

CELL MEMBRANE

CHLOROPLAST

Animal cell
An animal cell has no cell wall, just a cell membrane.
It lacks chloroplasts, but has other cell organelles
in common with eukaryotic cells.

Plant cell
A plant cell has a cell wall containing cellulose, and has
membrane-bound organelles called chloroplasts that
contain the light-capturing pigment chlorophyll. It also
has vacuoles which offer structural support and are
involved in storage and disposal of cell substances.

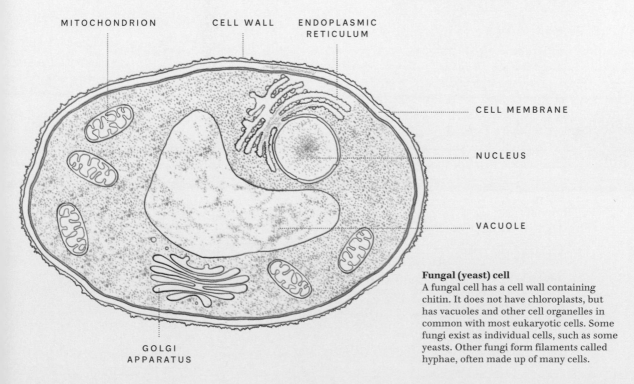

MITOCHONDRION   CELL WALL   ENDOPLASMIC
                                      RETICULUM

CELL MEMBRANE

NUCLEUS

VACUOLE

GOLGI
APPARATUS

Fungal (yeast) cell
A fungal cell has a cell wall containing
chitin. It does not have chloroplasts, but
has vacuoles and other cell organelles in
common with most eukaryotic cells. Some
fungi exist as individual cells, such as some
yeasts. Other fungi form filaments called
hyphae, often made up of many cells.

Fungi produce vast numbers of spores, often in exquisite reproductive structures, as a way of spreading to new habitats. Fungal reproduction – the process by which fungi produce spores – happens in two main ways: asexually and sexually.

FUNGAL REPRODUCTION

The structures that bear the spores of a fungus arise from the mycelium (the main body) of filamentous fungi (pp.36–37) or from fungal survival structures, such as ergots (pp.258–259). They can be microscopic, such as the spore-bearing structures of microfungi, or clearly visible to the naked eye, such as mushrooms.

A fungus can produce spores independently, without interacting with a compatible partner, in a process called asexual reproduction. Spores produced by asexual reproduction are both plentiful and genetically identical to the individual who produced them – rather like having lots of raffle tickets all with the same number.

Fungi also have the potential to mate with a partner of the same species in a process called sexual reproduction, whereby two compatible hyphae fuse and share genetic material. Sexual reproduction is more energy-intensive and produces fewer spores; however, each spore is different, containing a mixture of genetic material from both partners. This is like having fewer raffle tickets but all with different numbers.

Some fungi, such as the ascomycete blue mould *Penicillium expansum*, can reproduce only asexually, whereas others usually reproduce sexually, such as the mushroom *Agaricus bisporus* (pp.222–223). Some regularly reproduce both asexually and sexually. One example is the ascomycete plant pathogen *Pyrenopeziza brassicae*, which causes light leaf spot disease of oilseed rape and other brassicas.

WHY IS FUNGAL SEX IMPORTANT? If a fungus is growing in a stable (unchanging) environment, then producing lots of offspring that have identical characteristics to the parent fungus by the process of asexual reproduction is ideal – why change a winning formula? However, if the fungus is growing in a variable or more challenging environment – and if it can reproduce both asexually and sexually – it is much better for the fungus to hedge its bets by producing spores with many different combinations of characters, as achieved through sexual reproduction. Although more demanding and requiring liaison between two compatible fungal mates, spores produced sexually will have novel characteristics that will potentially offer the fungus a greater survival advantage.

Sexual cycle of *Mucor*
Species of *Mucor* fungus have two sexes. They are labelled + and - because they do not have any features that distinguish them as being male or female. Hyphae of different sexes (mating types) are attracted to each other. They mate when special structures called gametangia fuse together.

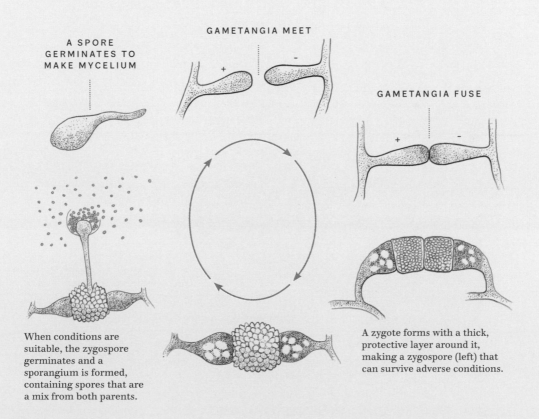

A SPORE GERMINATES TO MAKE MYCELIUM

GAMETANGIA MEET

GAMETANGIA FUSE

When conditions are suitable, the zygospore germinates and a sporangium is formed, containing spores that are a mix from both parents.

A zygote forms with a thick, protective layer around it, making a zygospore (left) that can survive adverse conditions.

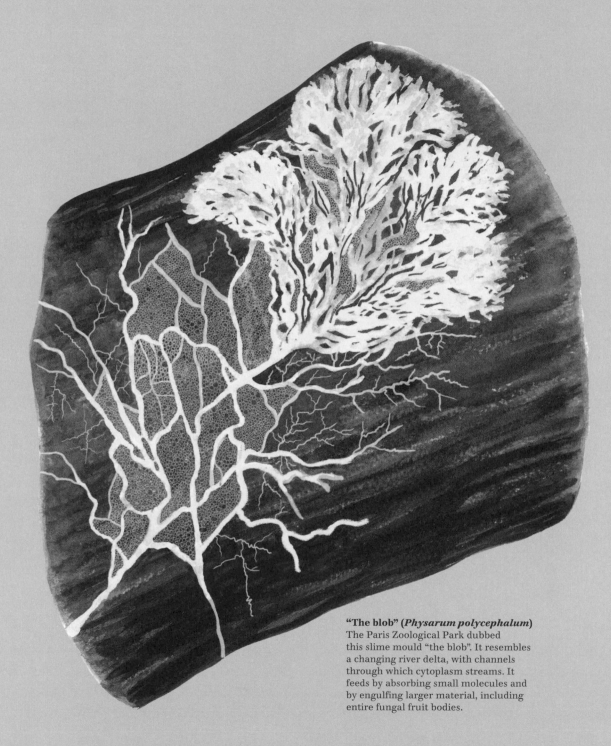

"The blob" (*Physarum polycephalum*)
The Paris Zoological Park dubbed
this slime mould "the blob". It resembles
a changing river delta, with channels
through which cytoplasm streams. It
feeds by absorbing small molecules and
by engulfing larger material, including
entire fungal fruit bodies.

Think you know a fungus when you see one? Think again.
Oomycetes and myxomycetes are often found in similar habitats
to fungi and work in similar ways – but they are
not even closely related.

SLIME MOULDS AND WATER MOULDS

OOMYCETES A type of water mould, oomycetes have cellulose in their cell walls, like plants. They are more closely related to diatoms (p.288) and brown algae than fungi, but, like fungi, they have hyphae which grow from the tip and branch to form colonies. They also reproduce by making spores. Some are decomposers (pp.38–39), while others are pathogens (p.289). Probably the most notorious oomycetes are *Phytophthora* species, which can cause devastating plant diseases. *Phytophthora ramorum*, for example, is wiping out huge areas of live oak in California and larch in Britain.

MYXOMYCETES The fruit bodies of some slime moulds look like tiny fungal fruit bodies, some on delicate stems that crumble when touched, others like a patch of yellow vomit. Myxomycetes have a complicated life cycle, assuming different body forms, including single-celled amoeba and cells that swim. The most remarkable is the change into a plasmodium – like a giant amoeba with many nuclei.

Though minute in some species, the yellow slimy mass of *Physarum polycephalum* can reach the size of a dinner plate. If food is spread out patchily, this myxomycete joins the patches, streaming nutrients between them. In a light-hearted experiment, scientists placed food in the pattern of Tokyo's underground stations, and the plasmodium linked the "stations" in a network that matched the railway.

POTATO BLIGHT

Phytophthora infestans causes potato late blight, blackening the foliage and rotting the tubers. This destructive disease of potatoes can be said to have changed the course of history. In the late 1840s, it destroyed most of Ireland's potato crop, causing the Great Famine. Over a million people starved to death and another million emigrated to North America.

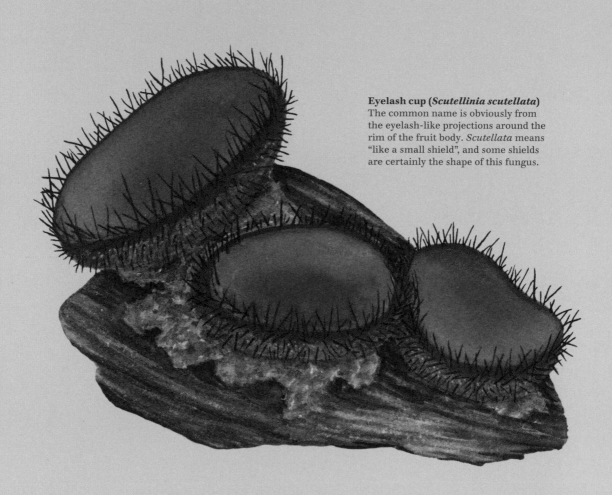

Eyelash cup (*Scutellinia scutellata*)
The common name is obviously from
the eyelash-like projections around the
rim of the fruit body. *Scutellata* means
"like a small shield", and some shields
are certainly the shape of this fungus.

WHY DO NAMES CHANGE?

Name changes reflect our changing understanding
of the evolutionary relationship between species.
Scientists' ability to sequence genomes, or "read"
DNA, allows us to work out which fungi are related.
Some fungi, which were previously thought to be
closely related because they look similar, have
turned out to be more closely related to other species
and so have been moved into another genus – hence
their genus name has changed.

As Shakespeare said, "a rose by any other name would smell as sweet". But names are important. Simply knowing an organism's name does not give us knowledge or understanding, but it does enable us to search for more information and to communicate with others about it.

WHAT'S IN A NAME?

HOW NAMES WORK Every organism has a scientific name, which usually derives from Latin or Greek and consists of two parts. That's why they're called binomial names ("bi" means "two"). Humans, for example, are *Homo sapiens*. The first part of the name indicates an organism's genus, in other words, which group of very closely related organisms it belongs to. The genus *Homo* included other species, such as the now-extinct *Homo neanderthalensis* and *Homo erectus*. The second word is the "species" part of the name, and it often describes a distinctive feature. Fungi are occasionally named after people, but you're not allowed to name a fungus after yourself.

COMMON NAMES Macroscopic fungi – that is, those visible to the naked eye – often have more memorable common names. But species may have different common names in different parts of the world. Even more confusing is when different species have the same common name. The ascomycete *Daldinia concentrica* goes by a variety of common names, including King Alfred's cakes (p.231), cramp balls, and the coal fungus. To avoid confusion, scientists use the binomial name written in *italic*, as each species has a single name that is recognized worldwide. For the same reason, we will use the English common name, when there is one, in this book, followed by the scientific name in brackets.

The more we look, the more fungal species we find. Present in just about every habitat on Earth, the true number of fungi that exist on our planet is still unknown and it may remain that way for many years to come.

HOW MANY FUNGI ARE THERE?

THE KNOWN KNOWNS Mycology became fashionable in Britain in the 1800s, when early mycologists began to formally record fungal finds. Much of the initial work on estimating the number of fungi on Earth came from the efforts of these pioneers, but the first predictions weren't made until the 20th century.

By the late 1900s, fungal estimates were based on the knowledge that, for every plant species, there were about six fungal species. At the time there were around 270,000 plant species known to science. Initially, many fungi were recorded twice – in both their sexual and asexual stage – and not all types of plant material that contain fungi were included in the count, such as inside plants and roots. With this in mind, scientists predicted that there were approximately 1.5 million species of fungi.

THE KNOWN UNKNOWNS By the 2000s, scientists realized that fungi do not solely associate with plants – they are everywhere! Advances in DNA technology enabled scientists to discover fungal signatures (genetic material known to be of fungal origin) in environments that had never been investigated before, such as in deep-sea sediments, rocks, and even inside humans. Furthermore, studies on known fungal communities showed that there are many more than can be physically cultured and identified. As such, the estimate changed to between 3.5 and 5 million fungal species.

THE UNKNOWN UNKNOWNS In recent years, it has been suggested that for every fungus that can be cultured there are eight that cannot be cultured. These "dark matter" fungi are only identifiable by their DNA signature. If this is true, there could be as many as 12 million fungal species on Earth.

With a predicted 90 per cent of all fungi still to be discovered, we are only at the tip of the iceberg. The more we discover, the more we learn – and from what we know so far, fungi play an important role in our lives and on our planet.

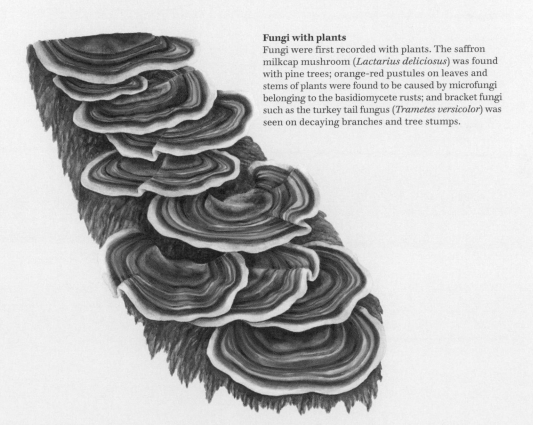

Fungi with plants
Fungi were first recorded with plants. The saffron milkcap mushroom (*Lactarius deliciosus*) was found with pine trees; orange-red pustules on leaves and stems of plants were found to be caused by microfungi belonging to the basidiomycete rusts; and bracket fungi such as the turkey tail fungus (*Trametes versicolor*) was seen on decaying branches and tree stumps.

We can't know exactly when fungi first appeared on Earth, but we do know they have been around for millennia – and the clues they've left paint an extraordinary picture.

ANCIENT FUNGI

2.4 BYA

Fungal structures don't tend to form good fossils, and the fine threads (hyphae) that fungi form are difficult to distinguish from other filamentous microbes. This makes it difficult to accurately pinpoint when fungi first appeared on Earth. The earliest signs are fungal-like filaments found in 2.4-billion-year-old rock – which would make fungi more ancient than originally thought (p.13).

635 MYA

Terrestrial fossil evidence indicates that fungi colonized land at least 635 million years ago. Early colonization of land by fungi can be explained in part by their filaments, which grow in search of food. These tough, thread-like structures can traverse gaps, penetrate rock crevices, and even break down rock – enabling them to live on land long before vegetation.

420 MYA

By the time primitive plants started to appear, many more fungal forms existed. Some of the most exquisitely preserved are found in the Rhynie Chert beds in Scotland, and show clear examples of how fungi associated with plants as mycorrhizal partners (pp.60–61) as well as plant pathogens. Evidence suggests that, before the first trees, the largest organisms were fibrous fungal structures, such as *Prototaxites*.

THE EVOLUTION OF FUNGI:
WHAT WE CAN LEARN FROM FOSSILS

PROTOTAXITES TOWERED OVER
OTHER ORGANISMS (420MYA)

251MYA

By the end of the Permian Era, conifers dominated the landscape. The fossil record from this period reveals the first definitive white-rot fungus. When the Permian Era came to a dramatic end with Earth's greatest extinction event, a fungal spike appears in the fossil record. This shows that, against the backdrop of mass extinctions, fungi not only survived but conquered.

120MYA

Mushroom fruit bodies were present in the fossil record. The oldest discovered to date is a 120-million-year-old gilled mushroom called *Gondwanagaricites magnificus*, which is beautifully preserved in a type of limestone.

52MYA

By the Eocene Era, mushrooms and other fruit bodies, mycorrhizal associations (pp.28–29), and even a fungal parasite growing from a fossilized ant and preserved in amber, give us a more detailed picture of life millions of years ago.

Fungi have partnered with plants and algae for millions of years. In fact, without fungi, plants might never have colonized land – and humans might never have existed.

ANCIENT PLANT PARTNERS

Fungi form mycorrhizal partnerships with plants (pp.60–61), and lichen partnerships with green algae and cyanobacteria (pp.70–71). Both partnerships require a fungal partner (the mycobiont) and one or more photosynthetic organisms as a partner (photobionts). Fossil evidence shows that these partnerships existed as long ago as the Devonian period, 420 million years ago.

Before then, most life existed in aquatic environments. Any life form attempting to colonize land needed protection from intense solar radiation and extreme temperatures, which made them prone to drying out. Soils lacked nutrients, and there was ten times more carbon dioxide in the atmosphere than there is today. So, to successfully live on the land, plants and algae needed to innovate – which is where fungi came in.

PARTNERING WITH PLANTS Two groups of fungi, the Glomeromycotina and the Mucoromycotina (pp.14–15), formed partnerships with ancestral plants. The then rootless plant partners benefited from association with fungal filaments (hyphae, p.16), which could extend into soil and sometimes rock, and which provided a large surface area, enabling the uptake of scarce nutrients and water. In return, plants shared with the fungi sugars made during photosynthesis. This formidable partnership was so successful that it is still in existence today with almost every plant on Earth, and is essential for our ecosystems to function.

As rootless plants evolved to form roots and shoots, so their valued partnerships with fungi evolved too. The fungi formed close associations with roots, extending and enhancing the benefits of the early associations. Once trees were around, a new group of fungal associations – the ectomycorrhizas – evolved. Both partnerships are maintained today (pp.60–63). More plant life meant more oxygen in the atmosphere, which ultimately led to the evolution of animals, including humans.

PARTNERING WITH ALGAE At around the same time but independently of plants, green algae (Chlorophyta) and some cyanobacteria began to form partnerships with fungi belonging to the phylum Ascomycota (p.15). The result was lichens (pp.70–71), which could survive the harsh environment of ancient Earth. The lichen body was made up of fungal hyphal filaments wrapped around algae, which offered the algae protection from desiccation and UV irradiation while also enabling the fungus and its partner to share vital resources, similar to mycorrhizal partnerships. Today, lichens are often present in extreme habitats, a visual reminder of their ancestral past.

Early plant forms
Aglaophyton major was an early rootless land plant of the Devonian period around 410 million years ago. It is one of the earliest land plants known to have had a mycorrhizal partnership with fungi, relying on fungal filaments to capture water and nutrients from primitive soils.

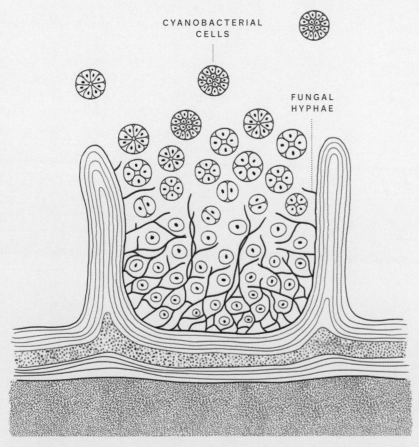

CYANOBACTERIAL
CELLS

FUNGAL
HYPHAE

A primitive lichen
Winfrenatia reticulata is the oldest recorded fossil lichen, discovered in fossils from the Devonian period in the Rhynie Chert, Scotland. It existed around 410 million years ago and comprised a layer of fungal hyphae (the thallus) with several depressions on its surface, each depression containing a net of filaments surrounding cyanobacteria.

*Fungi can be found in the **Arctic tundra** and in hot **deserts**, where there is little vegetation, as well as the **Antarctic**, where many fungi exist as lichens.*

Wherever there is a source of food and moisture, you will find a fungus. Fungi are everywhere: from the Arctic to the tropics, in water and on land, in the air and within plants and animals, and even in space.

WHERE ARE FUNGI FOUND?

You can find fungi in just about every habitat on Earth. But there are many more fungi in the outside environment than indoors.

OUTDOORS In gardens, fungi live in the soil, compost heaps, and log piles, where they play a vital role in decomposing dead plant and animal remains and breaking down dung (pp.144–145). Other fungi cause diseases, such as rusts, powdery mildews, and leaf spots (pp.72–77) on ornamental and vegetable plants. Many fungi are less conspicuous: endophytes (pp.68–69) live inside plant cells, while mycorrhizal fungi (pp.60–61) form beneficial partnerships with plant and tree roots. Some fungi can grow in unexpected places, such as caves, where they feed on dead animal and plant life (pp.158–159). They can even grow on food-scarce rocks, and sometimes cause damage to buildings and monuments. Rivers, streams, and oceans play host to fungi (pp.150–153). They even exist in and on our bodies (pp.92–93).

INDOORS Fungal spores stick to clothing and the soles of our shoes or on our pet's fur. In fact, most surfaces within the home will host fungi (pp.32–33).

Fungi are easily spotted in most of our homes at one time or another. They are usually decay fungi performing their important role of recycling – but, in this context, they are not a welcome sight.

AROUND THE HOME

Kitchens are the most common place to find fungi. Patches of asexual spores (pp.54–55) are produced by microfungi rotting our fruit and vegetables, cheese, bread, and jam. Fungi such as *Penicillium* and *Botrytis* appear as moulds on fruit, while *Mucor* is commonly found on bread and *Aspergillus* on jam. Species of black yeasts (*Exophiala*), red yeasts (*Rhodotorula*), and white yeasts (*Candida*) can even grow in the alkaline, high-temperature conditions of dishwashers.

Fungi such as *Cladosporium*, *Alternaria*, and *Stachybotrys chartarum* can grow on soft furnishings containing natural fibres, such as carpets, wallpaper, and books, where they appear as black moulds. You might also spot black moulds in bathrooms and other damp areas. These are a particular problem after flooding and produce airborne spores and toxic chemicals, so should be removed. Fungi such as the cellar fungus (*Coniophora puteana*) can grow in damp wood in buildings, causing brown rot (pp.120–121), which is crumbly and brown and quickly weakens the wood.

DRY ROT FUNGUS The dry rot fungus (*Serpula lacrymans* var. *lacrymans*) can start to grow in damp conditions. It is particularly troublesome as it then moves water through its mycelial cords (pp.118–119), enabling it to extend many metres through dry plaster and brickwork, and to feed on dry wood, causing a brown rot.

*The name **dry rot fungus** is a misnomer because, like all fungi, it needs water. It produces **prolific mycelium**, tell-tale brown, flat, spreading fruit bodies with a white margin, and rust-coloured spores.*

The fluted bird's nest fungus (Cyanthus striatus)

CHAPTER II

LIVE

THERE IS MORE TO A FUNGUS than meets the eye. Beneath each mushroom or fruit body is the mycelium: a vast network, from minute to enormous, below the ground or in living organisms or dead material, feeding, growing, communicating, and responding to the environment. Discover the life cycle of a fungus and learn about mycelium, fruit bodies, and spores.

Mycelium is key to the way most fungi live. Made up of a network of fine filaments called hyphae, the mycelium forms the main body of the fungus.

GROWTH AND NETWORKS

The hyphae that make up a mycelium are microscopic, up to 40 times narrower than a human hair. Their width is usually just a few micrometres (μm) – a few thousandths of a millimetre – and they can be as thin as 0.5 μm. They also usually grow away from each other, which spreads them out more thinly. However, in a few very significant basidiomycetes, hyphae grow towards each other and form cords and rhizomorphs (pp.118–119), which can be a millimetre or more in diameter. This makes them visible to the naked eye. Cords and rhizomorphs form networks that connect decaying wood they are feeding on or plants they are partnering with (pp.60–63). In fact, one of the largest organisms on the planet is a fungus (see left).

UNASSUMING BUT POWERFUL Mycelia can penetrate solid, bulky materials using physical force together with enzymes. They can also grow through narrow gaps and across empty spaces. Their large surface area (see right) is ideal for feeding – digesting food outside their bodies and then soaking it up (pp.38–39) – and makes them great partners for plant roots (pp.60–61). Just like plants grow from seeds, a new mycelium develops from a spore.

RECORD-BREAKING FUNGI

In North America, networks of rhizomorphs of individual honey fungi (*Armillaria*) are vast. The current record-breaker is in Oregon, USA, and covers 9.5 sq. km (3.7 sq. miles), weighs about 400,000 kg (880,000 lb), and is at least 2,500 years old.

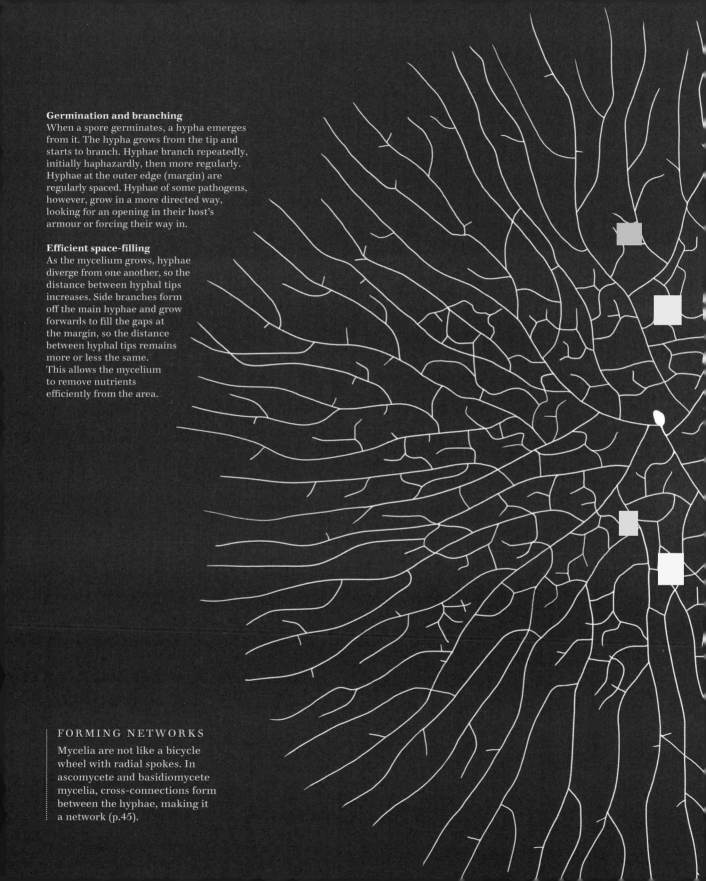

Germination and branching
When a spore germinates, a hypha emerges from it. The hypha grows from the tip and starts to branch. Hyphae branch repeatedly, initially haphazardly, then more regularly. Hyphae at the outer edge (margin) are regularly spaced. Hyphae of some pathogens, however, grow in a more directed way, looking for an opening in their host's armour or forcing their way in.

Efficient space-filling
As the mycelium grows, hyphae diverge from one another, so the distance between hyphal tips increases. Side branches form off the main hyphae and grow forwards to fill the gaps at the margin, so the distance between hyphal tips remains more or less the same. This allows the mycelium to remove nutrients efficiently from the area.

FORMING NETWORKS
Mycelia are not like a bicycle wheel with radial spokes. In ascomycete and basidiomycete mycelia, cross-connections form between the hyphae, making it a network (p.45).

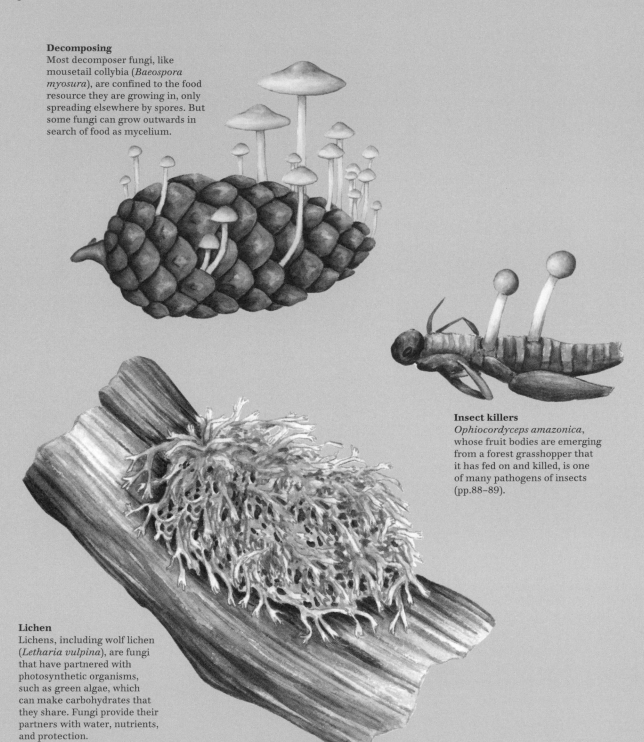

Decomposing
Most decomposer fungi, like
mousetail collybia (*Baeospora
myosura*), are confined to the food
resource they are growing in, only
spreading elsewhere by spores. But
some fungi can grow outwards in
search of food as mycelium.

Insect killers
Ophiocordyceps amazonica,
whose fruit bodies are emerging
from a forest grasshopper that
it has fed on and killed, is one
of many pathogens of insects
(pp.88–89).

Lichen
Lichens, including wolf lichen
(*Letharia vulpina*), are fungi
that have partnered with
photosynthetic organisms,
such as green algae, which
can make carbohydrates that
they share. Fungi provide their
partners with water, nutrients,
and protection.

When it comes to feeding, fungi are more like humans than plants. They cannot make their own food using energy from sunlight, so they must find food that is already available.

HOW FUNGI FEED

Fungi feed in two ways. Either the mycelium finds food and digests it, or a partner provides food for instant use (pp.60–61). Humans eat food and digest it in the gut, where enzymes break it down into small molecules that the body can use. Fungi digest food outside their bodies by secreting digestive enzymes. They then absorb the food molecules through their hyphae, whose thin size gives them a huge surface area. The enzymes at the tips of hyphae, and the physical force hyphae exert, enable them to bore into solid food, even wood.

Across the spectrum of kingdom fungi, there is an array of digestive enzymes. In fact, fungal enzymes can break down any compound produced naturally by plants and animals. Of course, not every fungus has the full complement of enzymes at their disposal, and so fungi vary in their ability to digest and use different foods.

FINDING FOOD Fungi obtain their food in three main ways: by decomposing dead material, killing organisms, or living with a host. Decomposers (called saprotrophs) feed on dead plant, animal, and microbial matter. This is either the whole organism when it dies, or parts of the organism that are shed during its life, such as leaves, branches, root cells lost as roots move through soil, faeces, skin cells, fur, or hair. Killers (called necrotrophs) kill tissues or the whole body of other organisms (pp.72–81, 88–91), which they feed on before, during, and after its death.

Fungi that feed off living hosts (called biotrophs) can be either harmful or helpful. Fungal parasites remove nutrients from the plant (pp.72–73), animal, or other fungus, limiting their growth or reproduction. Other fungi form mutually beneficial partnerships. For example, they partner with plant roots to form mycorrhizas (pp.60–61) or with algae to form lichens (pp.70–71).

All organisms, including fungi, can sense the environment around them – from light to sound, chemicals to temperature – and they respond accordingly.

MYCELIUM SENSES

Humans have five main senses for monitoring and responding to the environment. Fungi have equivalent senses, and others too. They can respond by growing – or, in the case of chytrids (p.14), by swimming – towards or away from chemicals, such as food or poisons. Hyphae can sense the presence of other hyphae close by, presumably from chemicals in the air or those diffusing in films of water.

SIGHT Fungi cannot see the way animals can. Many even spend most of their lives in darkness. But fungi do have a range of photoreceptors or "light detectors", which allow fruit bodies to grow towards the light. This gives them an advantage, as spores spread better above ground. Sunlight also regulates when fruit bodies are made by some fungi (p.53).

TOUCH For some fungi, touch is very important. When spores of fungi that cause rust diseases in plants (pp.72–73) germinate, the hyphae have to get inside the plant. They often enter through air openings on the underside of leaves, which they find by touch. In barley, for example, the leaf cells are elongated and lined up in rows with the narrow ends touching. Between these narrow ends are the air openings. The leaf cells also have a curved surface, so there are troughs between the rows of cells. The hyphae grow at right angles to the troughs by feeling their way up and down the slopes, as this gives them the best chance of finding an opening through which to enter.

HEARING Fungi cannot hear as we do, but they can respond to sound. Though this is still a new area of research, scientists have found many types of sound in the soil that fungi might respond to. It has recently been discovered, for example, that roots of some plants can locate water from its vibrations. Above ground, *Venturia inequalis* – which causes apple scab disease – responds to vibrations caused by raindrops landing on leaves.

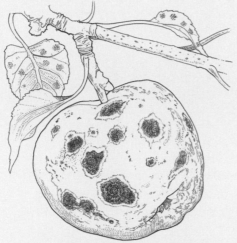

Touch: barley brown rust (*Puccinia hordei*)
This fungus can sense changes in the surface of barley leaves and produces infection structures when it encounters stomata (openings), thereby allowing the fungus easy entry into the plant.

Hearing: apple scab (*Venturia inequalis*)
The fungus overwinters in leaf litter on the orchard floor. In spring, minute flask-shaped fruit bodies emerge. The fungus senses vibrations caused by rainfall landing on the leaf surface, which triggers the fruit bodies to release their ascospores, allowing the fungus to spread to newly emerging leaves, blossoms, and fruits.

Sight: hat thrower (*Pilobolus*)
The hat thrower (p.145) has a lens-like system that allows it to detect light and grow towards it so it can shoot its packages of spores in that direction.

It was once thought that only animals could respond to encounters with other organisms or food. But we now know that organisms without brains or nervous systems can do so too – including fungi.

FUNGAL BEHAVIOUR AND MEMORY

Fungi can sense their environment and respond to it (pp.40–41), and they even have a sort of memory. Some of this happens at a microscopic scale, but scientists have been able to demonstrate it through simple experiments with cord-forming (pp.118–119) wood-decay basidiomycetes growing across soil in search of food.

BEHAVIOUR The mycelium of the sulphur tuft fungus (*Hypholoma fasciculare*) grows out of wood in search of food, at first in a roughly circular pattern. When it finds a new lump of food, it sends messages through its network to other searching parts of the mycelium, effectively telling it that food has been found. Searching in these other directions stops, and the unsuccessful mycelium probably reallocates much of its resources to the successful part. Hyphae grow towards each other and form a thick cord that joins the original and the new resource. Meanwhile, the rest of the mycelium disappears. When the fungus has fully colonized the new lump of food, it grows out on the other side and carries on searching. These foraging patterns look remarkably similar to the trails formed by termites.

How does the fungus send these messages? We don't know yet but, as with humans and other animals, it could use chemical messages and electrical signals. Indeed, electrical impulses have been detected in fungal mycelium and fruit bodies.

MEMORY Cord-forming fungi can also "remember" where the newly found food supply was relative to the position of the original. In an experiment on soil, the mycelium of *Phanerochaete velutina* was allowed to start to colonize a new food supply. The original block of wood was then cut away from the mycelium and placed on a new tray of soil. Mycelium grew out of this block again in search of food, but it grew much more extensively from the side of the wood that had originally been connected to the new food source.

HOW SOME FUNGI FORAGE FOR FOOD

Foraging for new resources
The mycelium of some decay fungi can grow out of the wood they are feeding on in search of fresh resources.

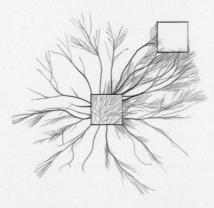

Responding to new food
When fungi find new sources of food, this is communicated to other parts of the mycelium, which then respond by growing more strongly towards the food source and disappearing from other regions. The mycelial architecture is remodelled.

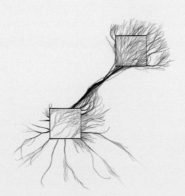

Connecting pipelines
A thick cord of tightly packed hyphae forms, connecting the original with the new resource. This is how large networks of cords form on the forest floor (pp.118–119). Nutrients and water can pass in both directions through these cords to parts of the mycelium where they are needed.

Mushrooms release thousands of spores, each one containing half the genetic material needed to make a new mushroom. But mushrooms are just the tip of the iceberg – the main body of the fungus is out of sight.

MUSHROOMS BELOW THE SURFACE

When a spore drops from a basidium – a club-shaped structure that protrudes from the gills (pp.174–175) on the underside of the mushroom cap – it is carried to a new location by the wind. If conditions are favourable, it will germinate, producing hyphae (fungal filaments) and eventually a mycelium (a network of hyphae; pp.36–37). If the fungus encounters a compatible mate, their hyphae will fuse and the two hyphae share their cell contents. One nucleus from each partner is maintained within the growing mycelium thanks to special hook-like structures called clamp connections. Mycelium often survives for many years in this state.

The mycelium extends into the soil, wood, or dead plants or animals to obtain food. When food is limited – or the environment, such as the temperature or light pattern, changes – the mycelium gears up for mushroom production. A knot of fungal hyphae forms a pinhead (primordia), an immature mushroom that begins to grow, producing a cap and gills containing many basidia. It is here, in each basidium, that the two nuclei – one from each partner – fuse and divide, usually producing four spores. When the mushroom is mature, the spores are released, and the process repeats. Meanwhile the parental mycelium can continue to grow and produce more fruit bodies.

MUSHROOMS BELOW THE SURFACE

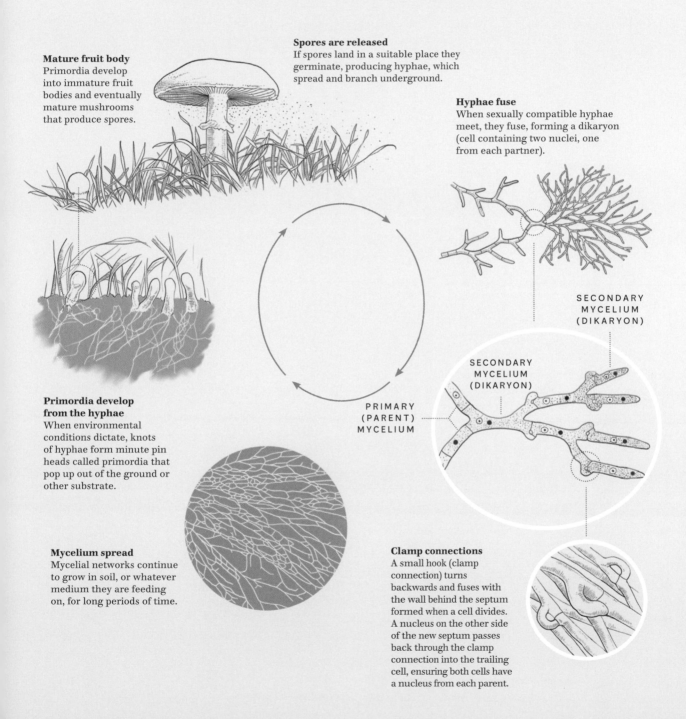

Mature fruit body
Primordia develop into immature fruit bodies and eventually mature mushrooms that produce spores.

Spores are released
If spores land in a suitable place they germinate, producing hyphae, which spread and branch underground.

Hyphae fuse
When sexually compatible hyphae meet, they fuse, forming a dikaryon (cell containing two nuclei, one from each partner).

SECONDARY
MYCELIUM
(DIKARYON)

SECONDARY
MYCELIUM
(DIKARYON)

PRIMARY
(PARENT)
MYCELIUM

Primordia develop from the hyphae
When environmental conditions dictate, knots of hyphae form minute pin heads called primordia that pop up out of the ground or other substrate.

Mycelium spread
Mycelial networks continue to grow in soil, or whatever medium they are feeding on, for long periods of time.

Clamp connections
A small hook (clamp connection) turns backwards and fuses with the wall behind the septum formed when a cell divides. A nucleus on the other side of the new septum passes back through the clamp connection into the trailing cell, ensuring both cells have a nucleus from each parent.

*There are over 52,000 species of basidiomycetes. Some
produce large fruit bodies, the most familiar of which
are mushrooms, while others are microscopic pathogens
of plants and animals.*

MUSHROOMS
AND OTHER
BASIDIOMYCETES

The role of fungal fruit bodies is to produce and release spores. Among basidiomycetes, the spores are produced on club-shaped cells (called basidia, hence basidiomycetes, pp.176–177), which are present on the fruit body's gills, on tube or pore walls, on teeth, or just on the surface of a crust.

Among basidiomycetes, the fleshy mushroom structure is the most iconic fruit body, but this is just one of the forms they can take. Some have stems, others can be tough crusts, or they might be bracket-/shelf-, or jelly-like. Colours range across the spectrum from off-white and dull brown to red and bright purple (pp.180–209).

These fruit bodies can be grouped based on what they look like. The names of these different fruit body groups are often based on the genus they resemble with "-oid", meaning "similar to", at the end. For example, clavarioids look similar to the genus *Clavaria*.

Bearded tooth fungus (*Hericium erinaceus*)
Many mushroom species produce their spores on gills, but "tooth fungi" make theirs on spines. This wood-decay fungus, which is rare in Europe, has spines that are 1–5 cm (⅓–2 in) long.

Beef steak (*Fistulina hepatica*)
This bracket fungus, with pores on the underside through which spores fall, has the look and consistency of raw meat when cut. It even "bleeds" red droplets. It is found mostly on oak and sweet chestnut tree trunks.

Violet coral (*Clavaria zollingeri*)
This violet-coloured, coral-shaped fruit body is 3–10 cm (1¼–4 in) tall. It is a rare fungus found in semi-natural (pp.140–141) grassland. These habitats are declining and its population is less than half of what it was 50 years ago.

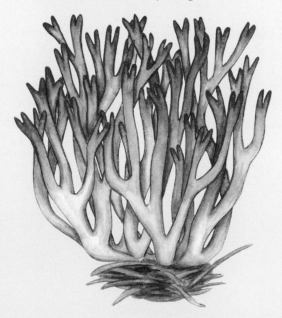

Apricot jelly (*Guepinia helvelloides*)
Also known as salmon salad, this fungus has fruit bodies that vary considerably in shape and size, between 4 and 10 cm (1½ and 10 in) tall. Its jelly-like or rubbery texture and salmon-pink or orange colour are striking. The basidia are mostly on the lower surface.

Yellow or common morel (*Morchella esculenta*)
There are over 80 species of morels, and they form the largest ascomycete fruit bodies, up to about 12 cm (4¾ in) tall. They are found in wooded areas and are mycorrhizal with tree roots (pp.60–61) or are rotters (pp.38–39). Some species are typically found on burnt ground. They can be confused with the deadly poisonous false morel lookalikes (pp.178–179).

Orange peel fungus (*Aleuria aurantia*)
Cup-shaped when young, the larger fruit bodies (up to 10 cm/4 in across) become twisted bowls that sometimes split. The asci form in the upper surface. They are rotters, particularly fruiting on disturbed ground in late summer or autumn.

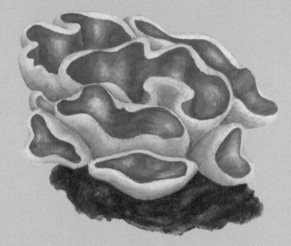

Candlesnuff fungus (*Xylaria hypoxylon*)
Found fruiting on branches and stumps of broadleaf trees, this fungus rots wood at the start of the decay process. The largely white fruit bodies produce asexual spores (conidia, pp.18–19, pp.54–55). In autumn and winter, these structures blacken and sexual spores (pp.54–55) are released from flask-shaped fruit bodies that are embedded within them.

Sac fungi, or ascomycetes, are the largest group of fungi. Over 97,000 species have already been described, and vast numbers are yet to be described by science. But relatively few sac fungi produce fruit bodies that are visible to the naked eye.

FLASKS, CUPS, SAUCERS, AND TRUFFLES

Some sac fungi are brightly coloured, but many are brown or black. Some are fleshy and occasionally have stems, whereas others are embedded in hard crusts or warty structures.

The difference between sac fungi and other types is how their spores are produced. Following mating, spores are produced in sacs (called asci, hence ascomycetes p.15). These are formed in the fruit bodies: cups, saucers, flasks, or truffles.

Flask fungi are so-called because their microscopic fruit bodies, which contain the sacs of spores, are flask-shaped with an opening at the top. Some are embedded in plant material and are too small to see with the naked eye. Others are arranged in larger fungal structures (stromata), which take many forms and are easily seen.

Cups and saucers are variations on a similar theme. As the names suggest, saucers are flatter, shallow shapes, whereas cups are taller and deeper. Sometimes raised above the ground on sturdy stems, cups and saucers can superficially resemble mushrooms, and the sacs are produced on the surface layer. In contrast, truffles are fruit bodies that grow underground as roundish dark lumps with a crusty exterior, with sacs present throughout.

All mushrooms grow in a similar way, following slightly different patterns – but first, the fungi have to mate.

HOW MUSHROOMS GROW

Most organisms have two sexes – male and female – but fungi that form mushrooms have hundreds. Usually, for a fruit body to form, two parent mycelia of different, compatible sexes must mate (pp.18–19). There is no visual difference between sexes, but fungi can differentiate sexes chemically. When mycelia mate, the hyphae simply fuse together. This only needs to occur once for the fungus to be able to produce fruit bodies when it has enough food and conditions are favourable (pp.44–45).

Mushrooms all form in similar ways, but there are three basic patterns depending on how the gills are protected when the fruit bodies emerge. The first two are the *Agaricus* and *Amanita* types (see right), and the third is the webcap (*Cortinarius*) type, in which the gills are protected by a veil of cobweb-like threads (p.171).

Agaricus type

When the tiny mushroom begins to form, a thin skin – the veil – joins the cap to the stem. The veil protects the gills as the mushroom pushes up through the soil. When it is close to full size, the veil breaks and part of it is often left on the stem, forming a ring. The spores can then drop from the exposed gills.

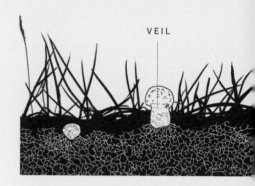

VEIL

Amanita type

Amanita has double protection. One veil protects the gills, and a second veil surrounds the whole fruit body as it emerges. When the veils break, one ring is left on the stem and another is left at the bulbous base called the volva. Little bits of veil are left on the cap, appearing as white flecks.

VEIL

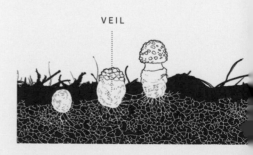

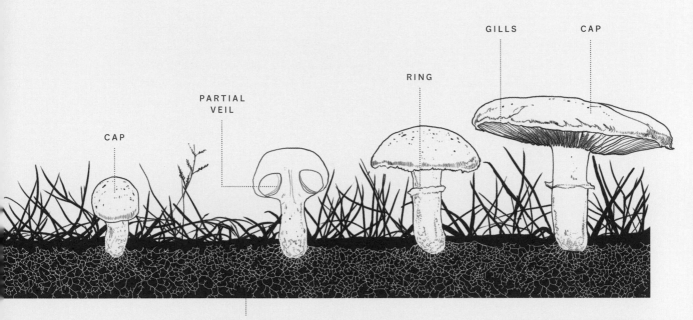

CAP

PARTIAL
VEIL

RING

GILLS

CAP

MYCELIA

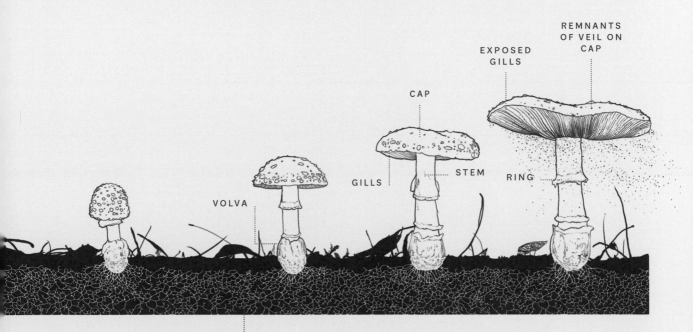

REMNANTS
OF VEIL ON
CAP

EXPOSED
GILLS

CAP

STEM

RING

GILLS

VOLVA

MYCELIA

Red-belted polypore (*Fomitopsis pinicola*) on a fallen beech trunk
Though rare in the UK this brown-rot wood-decay fungus is very common in mainland Europe on beech and conifers. The fruit body in the foreground formed after the tree fell and has vertical tubes, but the fungus further back is now on its side as it developed before the tree fell.

GRAVITY

The gills or tubes of fruit bodies must be vertical so spores can fall into the air currents beneath the fruit body. If a mushroom is moved so it is no longer upright, the stem turns and grows vertically again. When a bracket fungus is moved, it makes a new layer of tubes or a completely new fruit body with vertical tubes (see above).

*When fungi have sufficient energy and nutrients,
they have the potential to make fruit bodies. But this usually
only happens if environmental conditions are suitable.*

HOW THE ENVIRONMENT AFFECTS FRUIT BODIES

Light, temperature, and carbon dioxide – as well as water availability and humidity – all affect how or whether a fungus produces a fruit body, though their influence varies depending on the species. The exact impact of these environmental factors on fruit body formation has only been studied in a few commercially important species and some that produce fruit bodies quickly in the laboratory.

LIGHT AND TEMPERATURE Light increases fruiting in some species and decreases it in others. It can also affect whether fruit bodies are produced at all, and what they look like. Light stimulates production of fruit bodies of the velvet shank (*Flammulina velutipes*): the longer the light period, the more fruit bodies it produces. The ideal temperature for producing fruit bodies is usually several degrees lower than the ideal temperatures for mycelia to grow.

CARBON DIOXIDE The amount of carbon dioxide around mycelia affects how fruit bodies develop. In commercial cultivation of *Agaricus bisporus* (pp.222–223), carbon dioxide in the compost beds can rise to over 20 per cent. This is problematic as, above 1.5 per cent, stems grow long but caps are small, which buyers do not want. In nature, long stems have the advantage of raising fruit bodies higher into the air. Strangely, in the velvet shank, higher carbon dioxide reduces stem length.

Spores are the fungal equivalent of seeds in flowering plants. Invisible to the naked eye, these tiny packages of nutrients and genetic material are the main way fungi spread. They reach their final destinations in a variety of ways.

FUNGAL SPORES

All spores are microscopic, though we can see them if they fall en masse, for example, in a spore print (pp.172–173). Spores of different species vary in size, shape, colour, and ornamentation. They also survive for different lengths of time. A key difference is how they are formed. Some are only produced after mating and have a mix of characteristics from both parents (pp.44–45); others can be formed on hyphae even if mating has not occurred.

ASEXUAL SPORES Most asexual spores have thin walls, albeit often with sunblock pigments called melanins. Some – called chlamydospores – have very thick walls and are better able to survive inhospitable conditions. Although these spores are microscopic, you can spot some signs with the naked eye. The beautiful, branched formations of blue-green spores of the ascomycete *Penicillium*, for example, are visible in patches through blue cheeses. Meanwhile, Mucoromycotina pin moulds make their asexual spores (sporangiospores) in roughly spherical capsules, called sporangia, on the tips of hyphae that grow upwards. In *Rhizopus stolonifer*, a species of bread mould, sporangia can be seen as black dots borne on long hyphae.

SEXUAL SPORES The spores of fungi in different phyla (pp.14–15) form at different times after mating – some immediately, others after several months or even years. Some spores are adapted for survival, others for rapid spread. Those of Mucoromycotina fungi (pp.14–15) form immediately after mating. They are built for survival, with thick, often dark-coloured warty walls that enable them to survive longer while waiting for food; overarching branches from parent hyphae, which look like deer antlers, offer extra protection.

Ascospores – the sexual spores of ascomycetes – are made mostly in thin sacs called asci, which is why ascomycetes are sometimes called sac fungi. These sacs are found in cups, saucers, and flask-shaped fruit bodies (pp.48–49).

Basidiospores – the sexual spores of basidiomycetes – are made on club-shaped structures that are often quite broad, called basidia. They are produced on gills, tubes, and spines of mushrooms and other types of fruit bodies (pp.176–177).

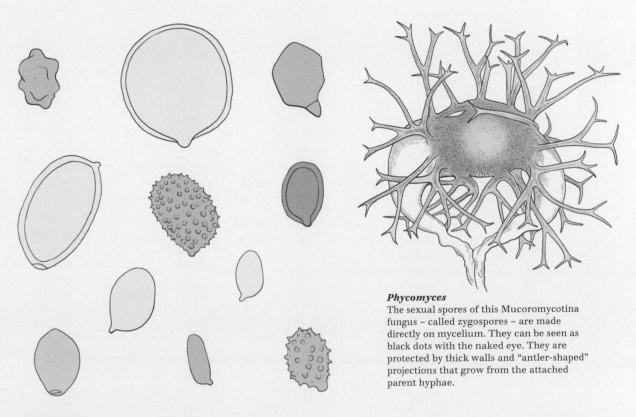

Phycomyces
The sexual spores of this Mucoromycotina fungus – called zygospores – are made directly on mycelium. They can be seen as black dots with the naked eye. They are protected by thick walls and "antler-shaped" projections that grow from the attached parent hyphae.

Basidiospores
The spores in basidiomycete fruit bodies are microscopic, often between five thousandths and twenty thousandths of a millimetre long. Their shape, colour, whether they have warty surfaces, and whether they contain starch also differs between species.

*Most spores rely on air or water currents
to move, and the majority land within just a few metres
of the fruit body that produced them. But some go
on much more exciting journeys.*

SPORE JOURNEYS

Only some types of spores – such as those of chytrids (pp.14–15), which can swim – can move themselves. For the rest, the chance of arriving in a suitable place with suitable conditions for growth is small. But fungi make millions of spores. A single field mushroom (*Agaricus campestris*) produces about 2.7 billion spores, which hugely increases the chances of success. Each year, over 50 million tonnes of spores are spread in the atmosphere, with vast clouds of spores forming above rainforests. In fact, it has been suggested that these might help rain clouds to form.

Occasionally, air currents can carry spores very long distances. The coffee leaf rust fungus (*Hemileia vastatrix*), for example, travelled from Africa to Brazil by wind. Spores can also be spread by raindrops or be actively "shot off" by the fungus itself. But a far less risky way for spores to arrive at their destination is to hitch a ride on an animal.

Raindrops

When raindrops land in the fruit body of the cup-like fluted bird's nest fungus (*Cyanthus striatus*), the fungus ejects spore "packages", which fly at 5 m (16 ft) per second. A tiny, coiled cord is attached to each package; when its free end touches a twig or leaf, the spore package sticks to it.

Touch

Puffballs and earthstars such as this collared earthstar (*Geastrum triplex*) make their spores in a spherical sac. A drop of rain, a falling twig, or the touch of a mammal causes enough pressure to force out a cloud of spores.

Hitching a lift

Spores are inadvertently carried on the bodies or in the mouths or guts of animals that feed on them, including mammals, insects, and slugs. The crinoline stinkhorn (*Phallus indusiatus*) produces smelly slime that attracts insects which then feed on it, inadvertently picking up spores and carrying them to new sites. Meanwhile, fungi that form mutual partnerships with insects are often taken by them to new locations (pp.82–83).

Russula olivacea

INTERACT

CHAPTER III

A FUNGUS RARELY keeps itself to itself. Fungi encounter other fungi, plants, animals, humans, and microbes – sometimes lending a helping hand, sometimes causing disease. From helping trees grow, to turning ants into zombies, to fighting each other, fungi can't help but interact with the world around them.

The vast majority of plants partner with fungi. The plant root plus
the fungus is called a mycorrhiza, from the Greek "mykes", meaning
fungus, and "rhiza", meaning root.

FUNGAL PARTNERSHIPS
WITH PLANTS

Mycorrhizal partnerships are an ancient way of life that originated around 450 million years ago. For most fungi involved, it is their only way of life. Both partners benefit from a mycorrhizal association: the plant gives the fungus sugars made during photosynthesis and, in return, the fungus gives the plant water and nutrients. Some mycorrhizal fungi help the plant tolerate adverse conditions, such as toxic chemicals, and the mycelium protects plant roots against some pathogens.

VARIETY OF PARTNERSHIPS There are many different types of mycorrhizas. The most common and widespread are arbuscular mycorrhizas: fungal hyphae grow in soil and penetrate young, fine roots, growing into the plant cells where they coil or branch into small, tree-like shapes where water, nutrients, and sugars are exchanged.

Most tree species in temperate and boreal latitudes, and tropical lowland rainforest trees form ectomycorrhizas. In these, the mycelium sheaths the fine absorptive roots like a fungal sock, and extends profusely into soil. The fungus grows into the roots, but does not penetrate into cells. Instead, it forms a hyphal network around the cells within the root, where exchanges occur.

Ericaceous plants, such as heather and rhododendrons, have their own mycorrhizas: the fungal hyphae grow in the outer cells of the fine hair roots. Up to 80 per cent of the volume of each of these plant cells is taken up with coiled hyphae, but it is still a mutual partnership.

*The roots of **90 per cent** of plants form mycorrhizal partnerships – probably the most frequent and **vital** close relationships between organisms from **different kingdoms**.*

Beneath every forest, underground networks connect trees and rotting vegetation. Information and resources pass through these networks. How are they formed? Fungi, of course.

WOOD WIDE WEB

Mycelia of mycorrhizal fungi (pp.60–61) and cord-forming wood-decay fungi (pp.118–119) grow away from their energy source in search of other food supplies, forming extensive, long-lived networks that are continually changing.

MYCORRHIZAL FUNGI Mycorrhizal fungi get their carbohydrate food directly from the trees they partner with, but they forage further afield for mineral nutrients and water (pp.60–61). Every tree has many different mycorrhizal fungi, and an individual fungus can form mycorrhizas with the roots of more than one tree, creating a network of several trees, sometimes of different species. Trees can be linked by several different fungal networks. These have been called the wood wide web. Nutrients, sugars, water, and chemical messages can move through fungal networks, but it is not certain how much movement there actually is between trees in forests. However, there is no doubt that mycorrhizal fungi are crucial to tree health.

WOOD-DECAY FUNGI Unlike mycorrhizal fungi, wood decayers have to find new wood to feed on because, as decomposers, they effectively eat themselves out of house and home. Some wood-decay fungi grow out of wood in search of new food. The mycelium joining the resources thickens into cords while other regions disappear (pp.42–43), eventually connecting lots of pieces of wood, forming another type of wood wide web. Honey fungi (*Armillaria* species) join dead and dying trees with their dark bootlace rhizomorphs (pp.118–119) in a similar way.

How the web protects species
As well as providing the plant with water and
mineral nutrients, it can provide protection
against some root-attacking pathogens. The
mycelial sock (pp.60–61) around small roots
not only provides a physical barrier but
mycelium also produces chemicals that
inhibit some microbes.

Ghost plants
So-called because they are white not green, ghost plants rely on underground fungi to obtain carbohydrates, water, and mineral nutrients, but they don't offer anything in return. They make use of the fungus's mycorrhizal relationships with trees to get sugars, making them double parasites.

Fungi are known for their parasitic behaviour, but they aren't the only organisms to cheat their partners out of nutrients. Some plants take advantage of the relationship between their roots and fungal mycelium.

PLANT CHEATERS

GHOST PLANTS *Monotropa* or ghost plants are a genus of flowering plants that cannot photosynthesize. Instead of obtaining energy from sunlight to make food, they obtain it via underground fungi. Ghost plants have fleshy root balls with few extensions into soil. The fungi grow into the roots and form a sheath around them, like ectomycorrhizas of trees (pp.60–61). The fungi simultaneously form mycorrhizal partnerships, allowing them to pass on sugars, which they obtain from the trees, to the ghost plants along with water and nutrients acquired from soil by the fungi. So, *Monotropa* parasitize not only the fungus but also the tree, using the fungus as a food transport pipeline.

ORCHIDS There are over 28,000 species of orchid. All have small seeds, which do not store much food. When they germinate, the orchid seeds send chemical signals to attract fungi, as they must rapidly encounter a fungal partner in order to survive. In all mutual mycorrhizal partnerships (pp.60–61), the fungus supplies the plant with mineral nutrients and water. In relationships with orchids, however, the fungus also supplies sugars, which it obtains from other sources. Some of the fungi involved are wood decayers, such as some species of *Trametes* and *Marasmius*, whereas others are plant pathogens, such as *Rhizoctonia*.

Most mature orchids have green leaves and can make their own sugars via photosynthesis, which they then share with the fungal partner. But more than 200 orchid species, such as the yellow-brown bird's nest orchid (*Neottia nidus-avis*), do not photosynthesize at all. This means they give nothing back to their fungal partner, effectively making these orchids parasites.

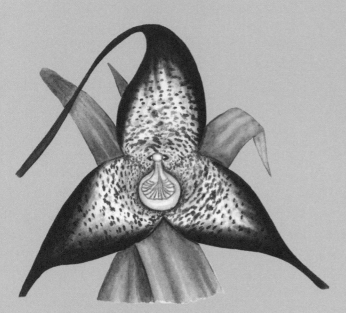

Cuckoo fungi

They might look different to termite eggs, but the smooth surface and chemical signals of the "balls" produced by *Athelia termitophila* fool the termites. Hyphae eventually grow out of the balls and consume the eggs.

Dracula orchids

The Dracula orchid smells like mushrooms, and its central petal (the labellum) actually looks like an upturned mushroom. This act of visual mimicry, along with the mushroomy smell, fools insects into visiting the flowers, pollinating them, and laying their eggs within the flower head.

Puccinia monoica

When this fungus infects Drummond's rockcress, it makes the plant form yellow "pseudoflowers". These "flowers" contain fungal reproductive structures that resemble buttercups, which are often growing nearby. By attracting insects, the "pseudoflowers" help to spread the fungus.

*Nature is full of tricksters. Some organisms cheat,
others mimic. Some even harm the organism they mimic.
Fungi are no different: some plants mimic fungi, and
some fungi mimic plants and even animals.*

MIMICS

MUSHROOMS OR FLOWERS? In the wet and misty cloud forests of Central and South America, mottled purple flowers with mushroom-shaped centres adorn the surface of trees. They even smell like mushrooms – but in fact they are Dracula orchids. The disguise attracts insect pollinators, which are searching for fungal fruit bodies in which they lay their eggs.

Fungi can deceive too. Some parasitic fungi alter a plant's appearance to help it spread. *Fusarium xyrophilum*, for example, is parasitic on yellow-eyed grasses in Guyana. The fungus stops the grass producing flowers. Instead, the fungal hyphae produce yellow, petal-like structures that look like the true plant flowers but produce spores. A similar strategy is deployed by *Puccinia monoica* (see left).

CUCKOO FUNGI *Termitomyces* fungi can form mutually beneficial partnerships with some groups of termites (pp.82–83). However, some fungi reap all the benefits and give nothing in return – in some cases, causing major harm. Fungi such as *Athelia termitophila* make survival structures, "termite balls", that are mistakenly collected by *Reticulitermes* and *Coptotermes* termites and taken to termite egg nurseries, where they are groomed and kept in a competitor-free, fairly constant environment – rather like cuckoos, who lay their eggs in the nests of other birds so they'll be fed by surrogate parents when the chicks hatch.

You wouldn't know from looking at them,
but all plants have microscopic organisms called
endophytes, usually types of bacteria or fungi,
living inside them. These hidden endophytes
can have a big impact.

HIDDEN FUNGI: ENDOPHYTES

PLANT CELL

HYPHAE

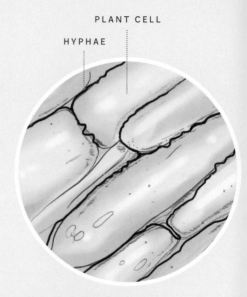

Some endophytes are decay fungi or pathogens waiting for the opportunity to feed, some are entirely incidental, and some provide benefits to the plant – for example, resistance to disease or by producing toxins that deter insects (see right). Most spread as spores and find their way into plants via natural openings or tiny wounds. Some spread via seeds of their host.

A HELPING HAND Dark, septate endophytes, so-called as they are dark-coloured and their cells have septa (p.16), have been found in roots of over 600 plant species, especially in harsh or stressful environments such as the Arctic, sandy soils, and sites contaminated with heavy metals. They probably help the plants tolerate these stresses and may help provide nutrients, as some form networks in roots akin to mycorrhizal fungi (pp.60–61).

Some ascomycetes and some basidiomycete fungi are found as endophytes in the roots, stems, and leaves of many herbaceous plants, especially in stressful environments. The microscopic *Fusarium culmorum*, for example, enables the coastal dune grass *Leymus mollis* to grow in saline soils.

A closer look at *Epichloë*
Epichloë grows as microscopic hyphae within grass shoots. The hyphae grow between the cells and do not penetrate them. Unlike most endophytes, *Epichloë* usually spreads by getting onto or into the grass seeds so when the plants grow from the seeds, the fungus is already present. They produce toxins including ergot alkaloids (pp.258–259).

Spore-producing structures of *Epichloë festucae* on shoots of the grass *Festuca rubra*
Epichloë festucae hyphae can also surround the flower head, forming a reproductive structure (stroma) and preventing further development of the grass. Flies, attracted by fungal compounds, deposit compatible fungal spores and sex occurs, forming a yellow pitted structure containing flask-shaped fruit bodies (pp.48–49), each with sacs containing sexual spores.

Sleepy grass (*Achnatherum robustum*)
Found in the western USA, sleepy grass contains endophytes that make LSD-like compounds (pp.258–259). When horses graze on it, they fall asleep – often for many days. Endophytes in drunken horse grass (*Achnatherum inebrians*) found in Asia, and in dronkgrasses (*Melica decumbens* and others) found in South Africa, have similar effects.

Lichens look like plants, but they are actually fungi that have teamed up with algae or cyanobacteria. The result is a whole new organism that can survive in some of the planet's most extreme environments.

LICHENS

There are about 20,000 known species of lichen. Most are ascomycetes but some are basidiomycetes. It's easy to tell the difference: basidiomycete lichens produce mushroom-shaped fruit bodies.

The fungi protect their partners and feed them water and nutrients, which they capture from the environment. In return, the algae and cyanobacteria give the fungus carbohydrates, which they make using energy from sunlight. At first glance, both parties seem to benefit. But since the fungi control the food – and hence growth – of their partners, it may well be more of a hostage situation.

THRIVING UNDER PRESSURE Most lichens grow very slowly: less than a millimetre each year. But because they can live for a long time, they can grow quite large. Some can be hundreds or even thousands of years old. Lichens commonly grow in stressful places, such as on exposed rock or trees, where neither the fungus nor alga could grow on their own.

Lichens are vital in extreme ecosystems: *Teloschistes capensis* dominates in deserts. The unusually fast-growing reindeer lichen (*Cladonia stellaris*) forms carpets over 15 cm (6 in) deep across vast areas of tundra and subarctic forests. It is the main winter food of reindeer, which can sniff it out through snow 1 m (3 ft) deep.

A. The most conspicuous lichens are fruticose. Attached to the substrate at a single point, they are made up of small, upright tubes that form tufts and often have a shabby appearance or dangle from branches. *Ramalina menziesii* is California's state lichen.

B. The most common lichens are crustose. These form thin crusts, growing in or on the substrate, and are not easily detached. They consist of a layer of algal cells covered by a distinct layer of fungal tissue. They grow slowly but are long-lived – some in the Arctic are thought to be over 1,000 years old.

A.

B.

C.

C. Common on trees and fallen logs in humid climates, foliose lichens are structured like a sandwich. The outside layers are a skin of tightly woven hyphae, inside which is a looser weft of hyphae, with algae cells just below the upper skin. On the underside is another tightly woven layer that attaches to whatever the lichen is growing on.

When fungi attack crops, it's not just the plants that are at risk. Fungal diseases can threaten the world's food production. Every year, they lay waste to hundreds of millions of tonnes of crops.

FUNGAL DISEASES AND CROPS

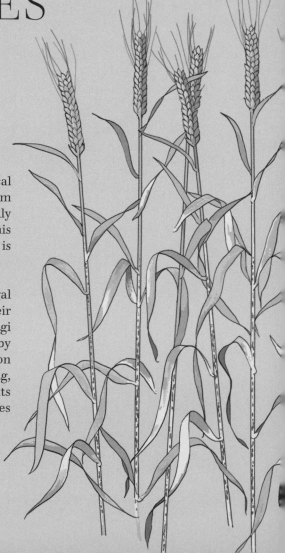

Most plants can survive fungal attack. They have physical barriers – such as bark on trees – that prevent the fungus from entering and causing infection. If this fails, plants can usually perceive the interloper and release chemicals that stop it. This is usually enough to prevent serious damage. But if a plant is susceptible to attack, the consequences can be devastating.

RUST DISEASES Rusts are among the most destructive fungal diseases of plants. A group of basidiomycete fungi, they get their name from the orange-coloured lesions they produce. These fungi have a fascinatingly complex life cycle. Black stem rust, caused by *Puccinia graminis*, infects wheat during its main growing season and then lies dormant in wheat debris over winter. In early spring, the fungus germinates, and spores spread to barberry plants (*Berberis vulgaris*), its alternate host. The fungus thus continues its life cycle until its preferred host becomes available again.

CORN SMUT When the basidiomycete *Ustilago maydis* infects maize, it produces black, tumour-like galls or smuts on the plant. Full of plant tissue, fungal filaments, and blueish-black spores, the smuts are edible and form the basis of the spongy, earthy-tasting Mexican delicacy *huitlacoche*.

GREY MOULD When soft fruits such as strawberries produce a fuzzy, grey-brown mould, the ascomycete *Botrytis cinerea* is usually the culprit. It can damage many crops – such as cucumbers and tomatoes – at different stages of the retail food chain. But grey mould can sometimes be beneficial. Under certain conditions the fungus causes noble rot of grapes, which are used to produce sweet dessert wines.

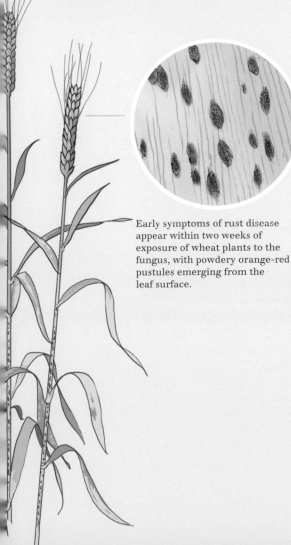

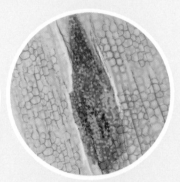

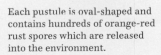

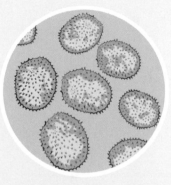

Early symptoms of rust disease appear within two weeks of exposure of wheat plants to the fungus, with powdery orange-red pustules emerging from the leaf surface.

Each pustule is oval-shaped and contains hundreds of orange-red rust spores which are released into the environment.

These warty spores are one of five different spore types produced by *Puccinia graminis* during its disease cycle on wheat and barberry.

*A devastating fungal disease is threatening the banana
as we know it. If scientists can't divert the crisis,
bananas could disappear from our fruit bowls altogether.*

SAVING THE BANANA

HOW THE BANANA CAME TO BE

Farmers propagated banana plants that bore fruits with fewer seeds, making them more palatable. But this eventually made the plants sterile, meaning they could no longer reproduce naturally. Today, banana plants are propagated like cuttings, so the new plant is a clone of the "parent". As it can't reproduce sexually, it can't develop genetic traits that might offer protection against fungal pathogens. This leaves modern banana plants very vulnerable.

The first banana plants were cultivated in Indonesia over 6,000 years ago. The main variety of banana became the Gros Michel, but by the end of the 1950s it had been completely wiped out. The culprit was a serious fungal disease called Panama disease or Fusarium wilt, caused by the pathogen *Fusarium oxysporum*. The fungus blocks the plant's water-carrying vessels, preventing water from reaching its leaves and causing the plant to wilt and eventually die.

The more resistant Cavendish variety, propagated from cuttings, was introduced as a replacement, and today it accounts for over 99 per cent of all exported bananas worldwide. A new variant of *Fusarium oxysporum* – more aggressive than the first – has spread through banana plantations and is now threatening the Cavendish variety.

Traditional fungicide treatments appear to be ineffective against this variant of *Fusarium oxysporum*. What's more, it can survive in the soil for decades without the presence of banana plants. Breeding resistant plant varieties is also problematic because all banana plants are related to the early Cavendish variety.

To save this well-loved fruit, scientists have turned to gene editing. The hope is that they will be able to manipulate the banana genome and propagate cells that are resistant to the fungus. Until then, the fate of the much-loved banana hangs in the balance.

HOW *FUSARIUM OXYSPORUM* INFECTS BANANA PLANTS

1.

The fungus can form thick-walled spores called chlamydospores which can survive in the soil for many years.

2.

Spores germinate in response to environmental cues such as nutrients released from the growing roots of banana plants.

3.

Fungal hyphae penetrate lateral roots and begin to grow between cells in the xylem (water-conducting tissue). At this stage there are few visible signs of fungal infection.

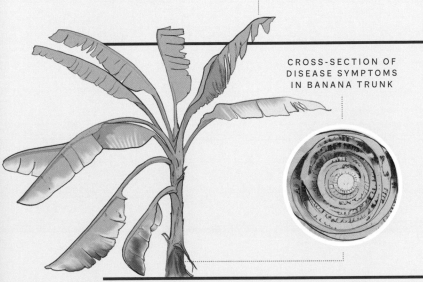

CROSS-SECTION OF DISEASE SYMPTOMS IN BANANA TRUNK

4.

The fungus blocks the movement of water, causing the plant to wilt and the lower leaves to begin to yellow.

5.

Eventually, the fungus reaches the leaves where it emerges and releases its spores.

6.

The disease is spread from plant to plant by overlapping roots and by infected leaves that fall and release fungal spores into the soil.

Just about every plant in the garden has a fungus associated with it. Many are beneficial mycorrhizas, others break down plant waste. But with over 70 per cent of plant diseases caused by fungi, they aren't always a gardener's friend.

THE GARDENER'S NIGHTMARE

HONEY FUNGUS There are several species of honey fungus (*Armillaria*), named after their honey-coloured mushrooms. With a faintly acidic aroma, the mushrooms are bioluminescent and glow in very dark conditions. Some of these fungi attack living trees and shrubs and feed off the dead wood.

Honey fungus spreads from one susceptible plant to another via mycelia of thick black or brown bootlace-like structures (called rhizomorphs, pp.118–119), which can spread over great distances. These fungi are a gardener's nightmare – not least because there is no way to control them other than digging up the infected plant root, all rhizomorphs, and other colonized material, and disposing of it, and then selecting plants that display some degree of resistance.

MILDEW AND ROTS Powdery mildew disease is caused by several fungi, forming a white, dusty coating on leaves, stems, and sometimes flowers. With a smaller area for photosynthesis, yield is severely affected. The host range of powdery mildew fungi is narrow. For instance, powdery mildew on *Acanthus* plants is caused by a different fungus to powdery mildew on cucumber. Unfortunately for gardeners, there are many different powdery mildews affecting many different plants.

A. Cherry leaf spot is widely known in the USA and Europe but became problematic in the UK in the 1990s as a result of international travel, including travel of plant material that wasn't sufficiently screened for the fungus. Caused by the fungus *Blumeriella jaapii*, the disease is recognized by purple spots on the cherry's leaves, which often turn yellow and fall prematurely.

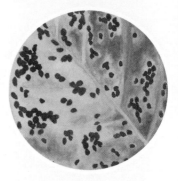

C. Any gardener with a rose plant will come across black spot, caused by the fungus *Diplocarpon rosae*. It produces black or purple spots on the upper side of leaves and can cause them to fall prematurely. Tar spot, caused by *Rhytisma acerinum*, is also fairly common, usually affecting sycamores (pp.122–123). The large black spots look a little unsightly but cause minimal damage.

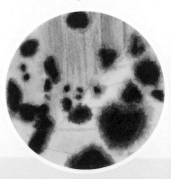

A.

B.

B. Cherries, apples, pears, and plums are susceptible to a range of rot fungi, especially *Rhizopus stolonifer*, *Penicillium expansum*, and *Monilinia* species. Brown rot, caused by *Monilinia*, is common in gardens and orchards in early autumn. Once the fungus has entered the fruit through a wound in the skin, it spreads rapidly through the flesh, causing it to fade and turn brown.

C.

D.

D. Coral spot, caused by *Nectria cinnabarina*, gets its name from its coral-pink pustules, which are full of spores. Its presence usually indicates that the host plant has underlying problems.

In many parts of the world, our landscape is dominated by trees. Over millennia, humans have altered this landscape continuously. Fungi, too, have changed it in very dramatic ways: by killing trees en masse.

ALTERING OUR GREEN LANDSCAPE

Humans have inadvertently moved fungal killers to new areas by trading seeds, living plants, and timber around the globe.

DUTCH ELM DISEASE Much of temperate Europe and North America was adorned with majestic elm trees until Dutch elm disease swept across the landscape. The first epidemic, caused by *Ophiostoma ulmi*, killed 10–40 per cent of elms in Europe from 1910 to the 1940s. A second, more destructive, wave began in the 1940s, caused by the related *Ophiostoma novo-ulmi*. This fungus makes toxins and blocks the tree's water transport system, causing it to wilt and die, killing millions of trees in the northern hemisphere.

CHESTNUT BLIGHT First seen on trees in the Bronx Zoo in New York in 1904, chestnut blight likely arrived in North America 30 years earlier on imported trees. It is caused by the ascomycete *Cryphonectria parasitica*, which enters a tree via a wound in the bark, kills living tissue, then forms sunken cankers that expand around the trunk, girdling the stem and killing the tree. American chestnut was the dominant tree species on the east coast but, by 1950, almost all had been killed, altering whole ecosystems.

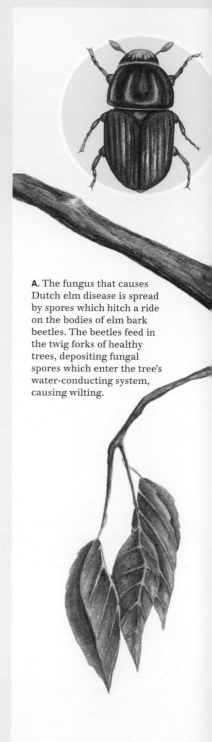

A. The fungus that causes Dutch elm disease is spread by spores which hitch a ride on the bodies of elm bark beetles. The beetles feed in the twig forks of healthy trees, depositing fungal spores which enter the tree's water-conducting system, causing wilting.

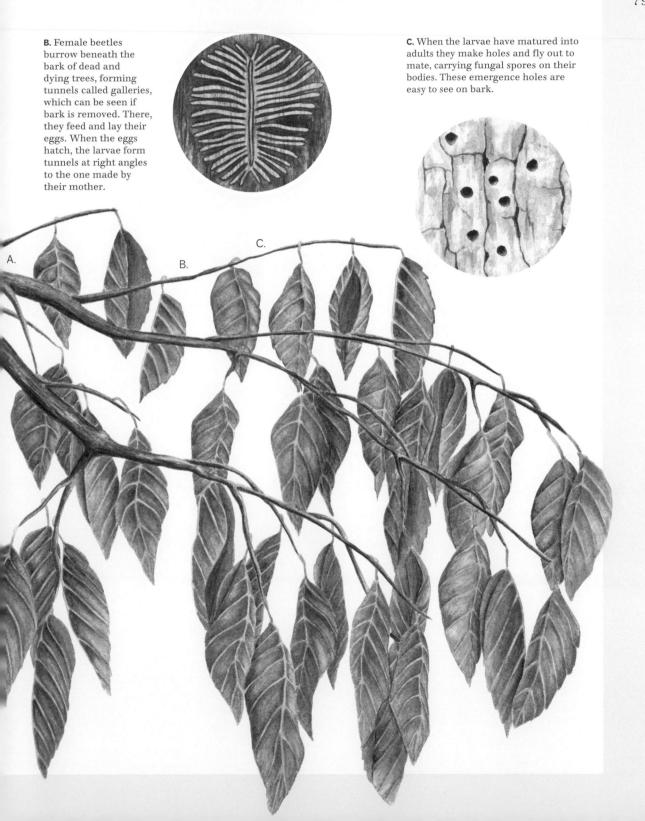

B. Female beetles burrow beneath the bark of dead and dying trees, forming tunnels called galleries, which can be seen if bark is removed. There, they feed and lay their eggs. When the eggs hatch, the larvae form tunnels at right angles to the one made by their mother.

C. When the larvae have matured into adults they make holes and fly out to mate, carrying fungal spores on their bodies. These emergence holes are easy to see on bark.

A.

B.

C.

In the last few decades, numerous new diseases caused by fungi have emerged, affecting a wide range of organisms from animals to ash trees.

EMERGING FUNGAL DISEASES

Fungal diseases of major crops and other plants (pp.72–77) are well known, but they are now threatening many more species, due mainly to changes in the climate and human activity. Fungi are now a recognized threat to species of bats, frogs (pp.90–91), bees, and loggerhead turtles. They also damage coral, causing sea fan aspergillosis (pp.152–153), and threaten certain flowering plants such as myrtle.

ASH DIEBACK DISEASE First observed in Poland in the early 1990s, ash dieback disease can kill whole trees across an entire landscape. This devastating disease is caused by the ascomycete *Hymenoscyphus fraxineus*, which is native to East Asia, where it does little harm to its host ash trees Manchurian ash and Chinese ash. The fungus can spread between forest nurseries on infected ash saplings and seed or via windborne sexual spores called ascospores (pp.54–55). It has now spread throughout much of Europe and is present in the UK and Ireland. The disease progresses more slowly in mature trees, and some are resistant, so there is hope that ash will not be completely lost.

ASH DIEBACK DISEASE CYCLE

Spores enter the leaf
Ascospores are spread
by wind and stick to
healthy leaves in the
canopy, where they
germinate and quickly
penetrate the leaf
surface. Fungal hyphae
grow within the leaf,
causing brown, dead
patches in late summer.

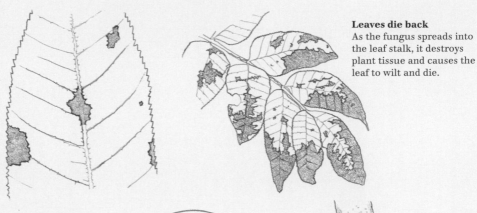

Leaves die back
As the fungus spreads into
the leaf stalk, it destroys
plant tissue and causes the
leaf to wilt and die.

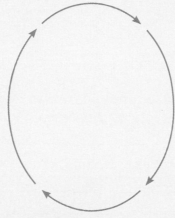

Mature fruit body
Cup-shaped fruit bodies appear in
summer. They open fully to release
ascospores in the mornings, coinciding
with dew fall, presumably to protect
the spores from drying out.

Ash dieback enters the stem
Leaves usually fall before the fungus
can grow into the stem, but if it does
grow from the leaf stalk into the stem,
lesions form and whole shoots can be
killed as the water-conducting tissues
are damaged. This dieback of shoots
is visible in summer.

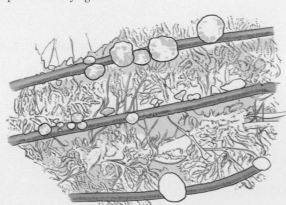

Leaf stalks harbour fungus
The fungus lies dormant in the stalks of fallen
leaves over the winter months, encasing the
leaf stalk in a resistant black layer that stops
other fungi from entering and maintains
a suitable environment until conditions are
right for fruit bodies to emerge.

Wood and leaves are difficult for insects to digest, so some have evolved mutually beneficial partnerships with fungi to solve this problem.

PARTNERING WITH INVERTEBRATES

Some insects bring food to the fungi, others take the fungi to food. The fungus then digests the plant material, retaining the nutrients in a form the invertebrates can feed on.

INVERTEBRATE FARMERS About 50 million years ago, leafcutter ants in South America became the first farmers when they started cultivating fungi in their nests. Or perhaps it was the other way around: fungi were the first organisms to have servants. The ants brought them food, kept them clean of pests, and provided them with a fairly constant environment.

HITCHING A LIFT Several thousand beetle species burrow beneath the bark of trees or even more deeply into the wood. Many of these so-called bark beetles carry fungi including blue-stain fungi, such as *Ophiostoma* and *Ceratocystis*, along with them. This is a win-win: the fungi are taken to a source of food, while certain beetle species consume the fungi. Sometimes, however, the tree loses out: some fungi and beetles end up killing their host tree.

Ambrosia beetles have special pouches to carry their partner fungus to a food source: dead, dying, and healthy trees. They burrow deeply, females pushing out displaced sawdust tubes or "noodles", and create tunnels or "galleries" where eggs are laid and larvae emerge. *Ambrosiella* and other fungal partners, which feed on the wood, grow on the walls of these galleries, giving both adults and their developing young larvae a more nutritious meal.

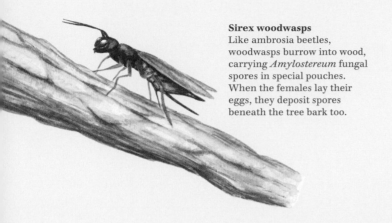

Sirex woodwasps

Like ambrosia beetles, woodwasps burrow into wood, carrying *Amylostereum* fungal spores in special pouches. When the females lay their eggs, they deposit spores beneath the tree bark too.

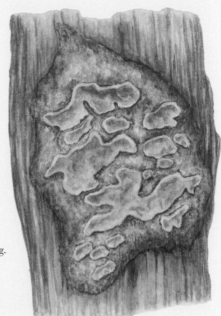

The crust fungus *Amylostereum*

Wood-decaying *Amylostereum* fungi soften the wood and improve its nutritional quality, allowing the woodwasp larvae to feed by burrowing.

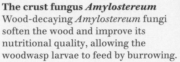

Termitomyces reticulatus

Similar to leafcutter ants, *Macrotermes* termites and *Termitomyces* fungi form a partnership that benefits them both. The worker termites bring plant material to their fungus "garden", in their nest below ground. The fungi digest the material and produce nutrient-rich balls of fungus, which the termites then eat.

Fungi partner with other organisms, like plants, in mutually beneficial relationships based on feeding, dispersal, providing a habitat, or a combination of these. They partner with birds and mammals in much the same way.

FUNGAL PARTNERSHIPS WITH BIRDS AND MAMMALS

With their remarkable range of enzymes that enable them to break down complex natural compounds, plus their ability to concentrate nutrients in low supply, fungi make an excellent source of food for many animals. But fungi first need to be able to get hold of – or be taken to – these complex food sources. Partnerships between fungi and animals have often developed based on these complementary needs (pp.82–83). Another important factor is the provision of a safe, stable habitat.

BIRDS Over 50 species of bird eat fungi and can thus spread spores over considerable distances. Australian king parrots, for example, devour the yellowy-orange *Cyttaria septentrionalis*. Hanging from tree branches in clusters, this fungus looks more like a plant fruit than an ascomycete fruit body.

At least 30 bird species associate with fungi to help them excavate nests in trees. The endangered red-cockaded woodpecker in North America, for example, prefers to excavate its nest in longleaf pine trees. These birds often opt for older trees, which have a softer heartwood (p.130) thanks to infection by fungi including *Porodaedalea pini*. Scientists have speculated that the birds could even facilitate the inoculation of the fungi into the trees in the first place.

HERBIVOROUS MAMMALS Herbivorous mammals eat plant material yet do not make all the enzymes necessary to break it down. They must therefore team with microbes inside the body, including fungi, to obtain energy and nutrients from their food. This system works for the fungi too: food is delivered to them, and they have a fairly stable environment.

For example, ruminating mammals – which include bison, cattle, and sheep – harbour fungi, bacteria, and other microbes in the rumen (effectively one of their stomachs). At 39–40.5°C (102–105°F), the rumen's temperature is higher than the mammal's body temperature and has hardly any oxygen. Instead, it is mostly carbon dioxide and methane. Unsuitable for most fungi, this environment is ideal for a group of chytrids (pp.14–15), namely Neocallimastigomycota, whose rhizoids penetrate the ingested plant material and help digest it.

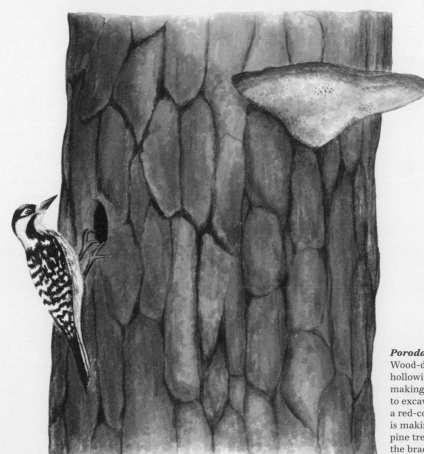

Porodaedalea pini
Wood-decay fungi cause hollowing and soften wood, making it easier for animals to excavate nest sites. Here a red-cockaded woodpecker is making a hole in a longleaf pine tree infected with the bracket fungus *Porodaedalea pini*.

Food for wildlife
Fungal fruit bodies are important sources of food for all sorts of wildlife. Many species of flies, beetles, slugs, and snails feed on them, and some even breed in them. Some invertebrates can safely feed on fungi that are poisonous to humans.

Fungi are highly nutritious, making them an attractive food source for many insects. But invertebrates provide a food source for some fungi, too. After all, what goes around comes around!

EAT OR BE EATEN

Many soil invertebrates, as well as some small mammals, eat fungal fruit bodies. Woodlice, millipedes, springtails, and flies eat mycelium growing in soil or inadvertently when eating rotting wood and leaves. Nematodes (roundworms) and some flies feed on mycelium, stems, and gills. On the other hand, Entomophthorales fungi and several groups of ascomycetes obtain their food by killing insects (pp.88–89). Some decay fungi in nutrient-scarce environments, especially those lacking nitrogen, trap and devour minute nematodes.

FUNGAL TRAPS More than 300 species of fungi have evolved elaborate traps to ensnare unsuspecting nematodes in soil. The simplest are tiny knobs on hyphae coated in a sticky material. When a nematode touches these lethal "lollipops" there is no escape; even if they break free, the knob is still attached. From this knob a hypha grows into the nematode and consumes it from the inside. Other fungi use a sticky loop or complex sticky network. Perhaps the most gruesome trap is the "noose" made of three cells, each of which immediately swells when a nematode passes through, the noose entrapping its prey. Other fungi produce spores that stick to the nematode, or that are hooked or curved in shape and stick in their throats. In all cases, hyphae grow inside and consume the nematode.

FRUIT BODIES AS FOOD Invertebrates often specialize in the type and part of fruit body they feed on. Some eat their way into the thicker parts, while others feed in the gills and pores. Nanosellinae beetles are tiny enough to climb into the tubes of bracket fungi, while some Staphylinid beetles brush off spores with their mouths, and Keroplatidae flies spin nets below brackets to catch falling spores.

Fungi that are pathogens of insects (entomopathogenic) often depend on their hosts for spore dispersal and employ macabre strategies to achieve this – including manipulating host behaviour and host mummification.

MUMMIES AND ZOMBIES

Over 1.5 million species of fungi are estimated to be predators of insects worldwide; some have even been employed as natural biological control agents (pp.278–279).

Many insects infected with entomopathogenic fungi, such as grasshoppers infected with *Entomophaga grylli*, unexpectedly climb and fix themselves to plant material in their final hours of life. The insect is positioned at an optimal height, giving the best environmental conditions for the dispersal of fungal spores. These symptoms are collectively known as "summit disease".

THE ZOMBIE-ANT FUNGUS Probably the most morbid summit disease is that of tropical carpenter ants infected with the ascomycete fungus known as the zombie-ant fungus (*Ophiocordyceps unilateralis*). Carpenter ants are mainly canopy-dwelling insects that descend to the forest floor only between breaks in the canopy. At ground level the ants follow defined trails that risk their exposure to infection by the fungus, which zombifies these ants for its own benefit, allowing the fungus to reproduce and spread its spores.

LIFE CYCLE OF THE ZOMBIE-ANT FUNGUS

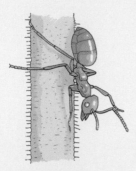

The death grip
It is within the fungal zone that the ant carries out its final act, biting into the underside of a sapling leaf, locking its jaws in a death grip. The ant is paralysed and dies of starvation

Mind-blowing effects
The ant travels back to the canopy to forage but within 2–4 days, the fungus begins to alter the behaviour of the infected ant. The ant moves downwards from the canopy and staggers around vegetation within the fungal zone, an area 25 cm (10 in) above ground level that is optimal for fungal growth.

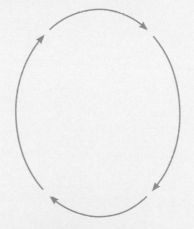

The fungus reproduces
The fungus continues to feed on the body content of the ant. A lollipop-like structure emerges from behind the head of the mummified ant carcass, enclosing minute flask-shaped fruit bodies containing sacs with sexual spores.

Fungal spores infect ant
An ant travelling between foraging sites, down to the ground from the canopy, passes through the killing zone. The spores stick to the ant's body and quickly penetrate the exoskeleton. Inside the ant's body, the fungus switches from a filamentous form to a yeast, and multiplies between the organs of the ant.

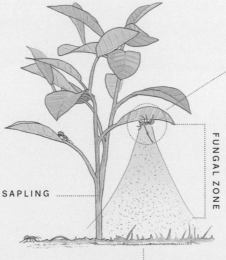

Spores are released
Thousands of spores are released, forming the killing zone, an area of forest floor where fungal spores have been deposited in the vicinity of passing carpenter ants.

SAPLING

FUNGAL ZONE

KILLING ZONE

*When fungi cause disease, the effects can be devastating.
For some animals, it can even mean extinction. Two emerging infectious
fungal diseases are causing widespread concern: chytridiomycosis
of amphibians and bat white-nose syndrome.*

AMPHIBIAN AND MAMMAL KILLERS

CHYTRIDIOMYCOSIS OF AMPHIBIANS In 1987, the golden frog became extinct in Costa Rica. The culprit? A chytrid fungus (pp.14–15) called *Batrachochytrium dendrobatidis*, known as Bd.

Chytrids have spores that swim in water, and this particular chytrid attaches to and burrows into the skin of amphibians. As these animals breathe and drink through their skin, the damage caused by the fungus disrupts their physiology and ultimately results in death.

Over 200 amphibians are now in decline because of Bd, and at least two species in addition to the golden frog have already become extinct: the gastric-brooding toad and sharp-snouted toad in Australia. The disease has spread rapidly and is now a worldwide problem.

A similar disease, caused by a related chytrid called *Batrachochytrium salamandrivorans*, has recently hit fire salamanders in Belgium and the Netherlands. At up to 35 cm (14 in) long, these are Europe's largest and most well-known salamanders, and they can usually live for more than 40 years. It's thought that this fungus, and Bd, have been introduced from elsewhere in the world.

BAT WHITE-NOSE SYNDROME Bats often hibernate for the winter in caves, where the temperature is 2ºC–14ºC (36ºF–57ºF) all year round. Once a safe haven, this habitat is often now shared with the cold-loving ascomycete *Pseudogymnoascus destructans*, which grows best between 5ºC and 10ºC (41ºF and 50ºF) but hardly at all above 15ºC (59ºF), matching exactly the conditions in the caves.

The fungus causes bat white-nose syndrome, which is having a devastating effect on bat populations in eastern North America and has now arrived in northern Europe. The disease gets its name from the white mycelium that grows on infected bats' noses, but it also appears on their ears and wing membranes. The skin tissues are eroded by the fungus, and the bats lose their fat reserves so cannot survive hibernation. White-nose syndrome affects many bat species, but the once very common little brown myotis has suffered to the point that it is facing local extinction.

The fungus that can **kill amphibians**, *known as Bd, has been spread partly by the* **human trade** *in these animals.*

Fungi on and in the human body
Every site on the body has resident fungi. The
diversity of fungi on the left and right sides of
the body are usually very similar, though it
often differs between individuals.

RESPIRATORY AIRWAYS

We breathe in, on average,
two million spores every day
through our nose and mouth.
The majority are common
airborne microfungi such as
Aspergillus and *Penicillium*
species, and fungal plant
pathogens such as *Cladosporium*
and *Alternaria* species.

LUNGS

In healthy individuals, fungal
spores that manage to reach the
lungs are quickly disposed of by
our body's immunity cells. Fungi
commonly found in lung tissue
include ascomycete yeasts such
as *Candida* species and
basidiomycete yeasts such as
Cryptococcus neoformans,
as well as common airborne
microfungi.

GUT

More than half of the fungal
species found on the skin, mouth
and lungs are also present in the
gut, suggesting that a common
route for transfer exists between
these body sites.

SKIN

There are certain fungi that are
skin specialists such as *Candida*,
which cannot survive for long
periods away from the human
host. *Malassezia* species, found
predominantly on the scalp, are
fungi that can no longer make
their own lipids and therefore
rely on obtaining these from
human sweat glands.

FEET

Feet host the most fungi, with
over 60 different types living
between the toes.

A diverse range of microscopic fungi live in and on humans – no human is fungus free. Each body has its own community of fungal residents, collectively known as the mycobiome.

THE HUMAN MYCOBIOME

You might have heard of the microbiome, the microorganisms (bacteria, fungi, and viruses) that live in and on our bodies. The mycobiome refers specifically to the fungi in the microbiome.

FUNGI ON THE BODY The fungi that commonly associate with humans are those that are routinely found in the environment. The ascomycete yeasts such as *Candida* species are common body residents, whereas the filamentous ascomycete fungi such as *Aspergillus* and *Penicillium*, and the basidiomycete yeast *Cryptococcus*, all have airborne spores which can be breathed in.

FROM BIRTH TO ADULTHOOD As we age, the diversity of fungi associated with us changes. This is an area of active research, but it is clear that most fungi and other microbes that live with us are vital to our health and well-being. Newborns inherit microorganisms, including fungi, from their mothers and from the environment. A baby born vaginally has a different range of fungi to one born by caesarean section. Breastfed babies receive additional fungi from their mothers. These early encounters with fungi and other microbes help develop immunity.

The skin-dwelling fungus *Malassezia* feeds off the oils in human sweat glands. At puberty we produce more oils and levels of this fungus increase. It breaks down oils and releases irritants, drying the scalp and causing dandruff. Many antidandruff shampoos contain antifungal agents to target *Malassezia*. As we age, we are more susceptible to oral yeast infections because we may produce less saliva. Our gut fungi also change, which may be why we become more prone to some gut conditions.

Fungi produce millions of offspring, making them one of the most reproductively active organisms on the planet. This also means the air contains spores all year round. Breathing in a few fungal spores is usually harmless, but it can trigger an allergic reaction for some.

FUNGAL ALLERGENS

INVISIBLE ALLERGENS Fungal spores are microscopic, so they cannot usually be seen unless you puff a puffball (p.57) or turn a mushroom cap-down to make a spore print (pp.172–73). Due to their size, spores are easily carried by air currents and can enter our respiratory airways. For most people, this causes no problem. However, fungal spores can be an issue for those with asthma, who are immunocompromised, or who have allergies such as hay fever.

THE MAIN CULPRITS Fungal allergens can cause symptoms such as a runny nose, itchy throat, fever, and headaches. The fungi most likely to trigger a reaction are microfungi, some of which are plant pathogens, such as *Alternaria*, *Cladosporium*, and *Didymella*. In Europe, peak spore release for these fungi coincides with the crop harvest in July through to September, while fungi such as *Aspergillus* and *Penicillium* – which are found inside as well as outside – release spores throughout the year.

Many of these fungi can survive on a wide range of materials: outside in compost heaps and in the soil, and in the home on shower curtains, windowsills, soft furnishings, or foodstuffs. In general, roughly half as many fungal spores are present inside compared with outside. However, if homes become damp or flooded, black moulds such as *Ulocladium chartarum* and *Stachybotrys chartarum* can be troublesome for allergy sufferers, though these are generally relatively easy to keep at bay by wiping with a mild detergent.

Air-borne spores
It is estimated that between 1,000 and 10,000 fungal spores are present in every cubic metre (35 cubic feet) of air – that's 100 times more than plant pollen. We breathe in about ten spores with every breath.

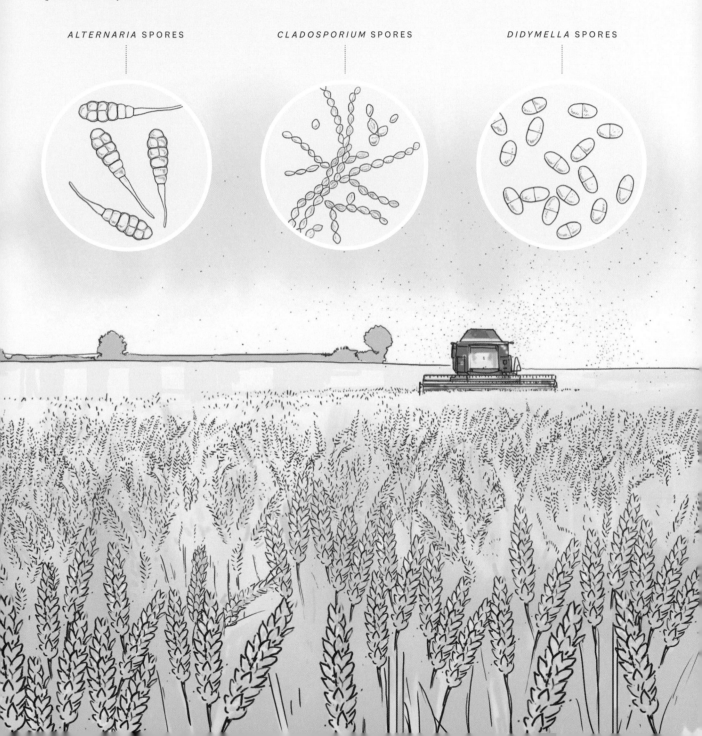

ALTERNARIA SPORES

CLADOSPORIUM SPORES

DIDYMELLA SPORES

**Deadly dapperling
(*Lepiota brunneoincarnata*)**
A rather fleshy mushroom with
a short stem about 2–5 cm (¾–2 in)
long, deadly dapperling contains
amatoxins. It has dark red-brown
scales towards the lower end of the
stem and a red-brown mottled cap.

POISONOUS MUSHROOMS

Some of the most common deadly mushrooms include:
Death cap (*Amanita phalloides*)
Destroying angel (*Amanita virosa*)
Funeral bell or autumn skullcap (*Galerina marginata*)
Deadly dapperling (*Lepiota brunneoincarnata*)
Common conecap (*Conocybe rugosa*)
Deadly webcap (*Cortinarius rubellus*)
False morel (*Gyromitra esculenta*)

Common conecap (*Conocybe rugosa*)
Growing on woodchips, in lawns, and in
compost, this deadly poisonous mushroom
is another species containing amatoxins. It
has a wrinkled orange-brown cap and a ring
that often turns rusty brown as it becomes
covered in spore deposits.

Many fungi produce toxins in their mycelium
or fruit bodies to deter competitors. Unfortunately,
these toxins also affect humans – some fatally.

FUNGAL TOXINS

Humans may be exposed to toxins in foods contaminated with certain fungi, or by ingesting a poisonous mushroom lookalike (pp.178–179). Microfungi produce many spores that are present in the environment and can grow on grain and other foods if moisture is present. Some produce toxins (mycotoxins) as defence chemicals to ward off competitors; these toxins can pose a serious threat to human and animal health if they are eaten. There are many mycotoxins, but those produced by *Aspergillus flavus* and *A. parasiticus* – called aflatoxins – are the main culprits of poisoning. Of the five main aflatoxins, aflatoxin B1 is the strongest. It enters the liver and is converted into a more potent toxin, which can cause liver disease and cancer.

POISONOUS MUSHROOMS Knowledge of poisonous mushrooms has been communicated between humans from one generation to the next. Today, we can identify and characterize the toxic chemicals in poisonous mushrooms and study their biological effects. The most poisonous mushrooms are those that produce toxins that cause cell damage. These include cyclopeptides (amatoxins and phallotoxins), orellanine, and gyromitrin.

Cyclopeptides and orellanine are both heat stable, meaning they are not destroyed by cooking. Allegedly, Agrippina, the fourth wife of the Roman emperor Claudius, used this knowledge to poison and kill her husband, allowing her biological son, Nero, to succeed his stepfather to the throne. She is said to have mixed the edible Caesar's mushroom (*Amanita caesarea*) with the death cap (*Amanita phalloides*) before cooking and serving them for her husband's supper.

*To healthy humans, fungi are more of a nuisance than
a threat, capable of causing superficial, easily managed
infections. But if the immune system is weakened, fungi can
take advantage – with serious, sometimes fatal, consequences.*

HUMAN DISEASES CAUSED BY FUNGI

Fungi are opportunistic organisms, and for people who are immunocompromised – for example, those with AIDS or undergoing chemotherapy – fungi can cause secondary infections that can be fatal.

CANDIDA The ascomycete yeast *Candida albicans* is a fungus shape-shifter: it can survive as a single-celled yeast and also as fine filaments. It is most problematic in its filamentous form. It colonizes the body, in particular the mucous membranes, causing conditions such as thrush, a common infection easily treated in healthy individuals, where tissues become covered in creamy speckles that resemble a thrush's chest. But the potentially life-threatening invasive candidiasis can occur if the immune system is compromised and *Candida* cells enter the bloodstream, where its filamentous forms can infect tissues, including the vital organs.

CRYPTOCOCCUS NEOFORMANS Cryptococcosis is an infection caused by the basidiomycete yeasts *Cryptococcus neoformans* and *C. gattii*, present in bird droppings and on fruit and bark of various trees. *C. neoformans* is found worldwide, *C. gattii* mainly in tropical and subtropical areas. These yeast cells are surrounded by a polysaccharide capsule protecting them from our body's defence system. In immunocompromised people, they can infect lung tissue, causing pneumonia-like symptoms called cryptococcosis. In more severe infections, the yeast hides in immunity cells called macrophages, using them as a "Trojan horse" to go through the bloodstream. They can cross the blood–brain barrier, where the yeasts enter brain tissue, causing life-threatening cryptococcal meningitis.

*The human body's **immune system**, when healthy, is well equipped to keep disease-causing fungi at bay.*

Fungi form communities and share habitats with other microorganisms, whether in the soil, on a particular food, or as part of our microbiome.

FUNGI AND BACTERIA

Fungi associate with bacteria in almost every ecosystem on Earth and play an important role in how these ecosystems work. Relationships can be intimate or less formal, beneficial or detrimental. Often, different kinds of fungi interact with different kinds of bacteria in the same ecosystem, and the impact of these partnerships is affected by various factors including: physical association (whether the bacterium is freely associating with the fungus, is part of a biofilm (p.13) around fungal hyphae, or is living inside the fungal cells); chemical signalling or "dialogue" between organisms; environmental conditions, such as pH; and the activity of the host (whether living or dead). Fungi also have resident bacteria just like we do. When thought of in this context, our own microbiomes are inhabited by fungi which themselves have microbiomes.

HYPHAL SUPERHIGHWAY Fungal hyphae form interconnections and can spread over large distances (pp.36–37). Each hypha is surrounded by a film of water, and some motile soil bacteria use this hyphal water jacket as a superhighway to swim along the hyphae, covering greater distances and moving across air pockets in the soil. Some partnerships benefit the fungus too. Phosphate-solubilizing bacteria, for example, are thought to help arbuscular mycorrhizal fungi to release phosphorus locked up in dead plant remains in a form that can be used by plant and fungus. Nitrogen-fixing bacteria such as *Kosakonia radicincitans* make nitrogen more available in the soil.

FUNGAL FRIENDS OR FUNGAL FOES?

1.

In the soil around the roots of a plant (the "rhizosphere") and often in the roots themselves, communities of fungi and bacteria exist. These microbes interact with one another, and the outcome of these interactions influences both the diversity of microbes present and the resulting vigour of the plant.

2.

The mycorrhizal-like endophyte *Serendipita indica* is a fungus beneficial to plants that lives within plant roots and the surrounding soil. The fungus is known to increase plant growth and protects it against certain environmental stresses.

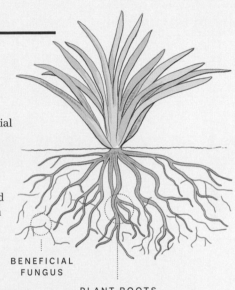

BENEFICIAL
FUNGUS

PLANT ROOTS

3.

Living alongside these beneficial fungi are bacteria that feed off them. The bacteria are either fungal killers (such as *Collimonas fungivorans*, below left), punching holes in the fungal cell wall and feeding off the contents, or they are fungal feeders (such as *Kosakonia radicincitans*, below right), living in close association with the hyphae, feeding off metabolites produced by the fungus.

4.

Bacteria that feed on fungal exudates colonize the fungal hyphal filaments, creating a biofilm of bacteria and slime – a protective jacket – that prevents the fungal-killing bacteria from reaching their fungal prey.

*COLLIMONAS
FUNGIVORANS*
- FUNGAL KILLERS

*KOSAKONIA
RADICINCITANS*
- FUNGAL FEEDERS

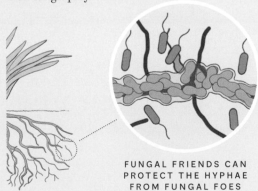

FUNGAL FRIENDS CAN
PROTECT THE HYPHAE
FROM FUNGAL FOES

Almost everywhere a fungus grows there will be other fungi and bacteria nearby that compete with it for space and nutrients. This means fungi are often on the warpath – and they have an arsenal of weapons at their disposal.

FUNGUS WARS

Among the defence and attack weapons of a fungus are enzymes, antibiotics (pp.252–253), and other toxic and inhibitory chemicals. Some of these chemicals are volatile gases, which spread to nearby microbes through the air; others spread through water in tiny pores in wood or soil. Fungi can even change the acidity of the environment to favour their own growth.

MYCELIAL BATTLES The mycelia of wood-decay basidiomycetes often meet over a wide front and bring to bear a formidable array of attack and defence chemicals. Victorious fungi kill their opponent and capture its territory, but battles can also end as a deadlock in which neither fungus takes territory from the other.

MYCOPARASITISM A few fungi feed by parasitizing other fungal species. Hyphae of the parasite lie on the surface or coil around hyphae of the fungus on which they are feeding. Some obtain food directly through the hyphal wall; in others, a thin hypha of the mycoparasite grows inside the other to take nutrients.

While parasitism is the only way of feeding for some fungi, others are temporarily parasitic, eventually killing their host, taking its territory, such as part of a log, and continuing to feed there.

A. The powdercap strangler (*Squamanita paradoxa*), with its brownish-violet cap, takes its parasitism one step further: it replaces the upper stem, cap, and gills of its victim, the earthy powdercap (*Cystoderma amianthinum*), with its own.

A.

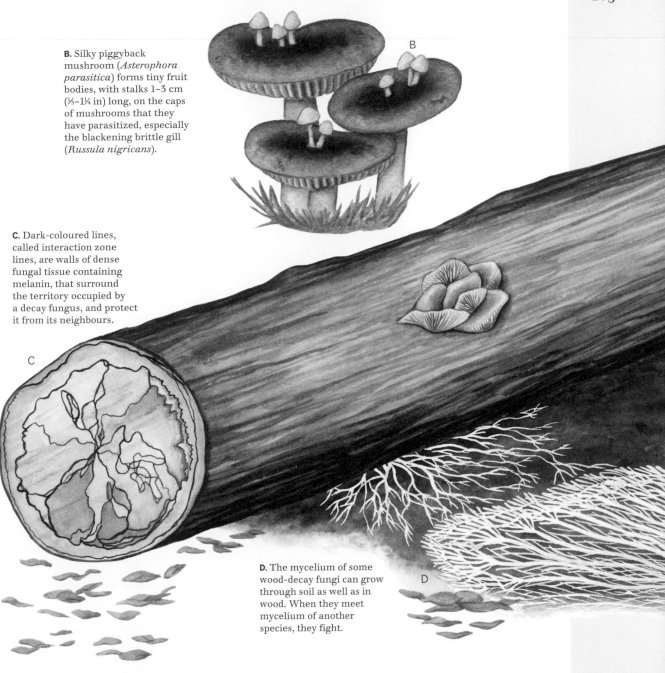

B. Silky piggyback mushroom (*Asterophora parasitica*) forms tiny fruit bodies, with stalks 1–3 cm (⅓–1¼ in) long, on the caps of mushrooms that they have parasitized, especially the blackening brittle gill (*Russula nigricans*).

C. Dark-coloured lines, called interaction zone lines, are walls of dense fungal tissue containing melanin, that surround the territory occupied by a decay fungus, and protect it from its neighbours.

D. The mycelium of some wood-decay fungi can grow through soil as well as in wood. When they meet mycelium of another species, they fight.

Devil's fingers (*Clathrus archeri*)

CHANGE

FUNGI HAVE LIVED ON EARTH for hundreds of millions of years, but the world is changing and stable habitats are becoming a thing of the past. How are fungi adapting to warmer climates? Like all living organisms, members of the fungal kingdom are under threat – but many are finding ways to adapt and even thrive in new habitats.

Our climate is changing, affecting the water cycle and rain: both where it falls and how much. Alongside elevated temperatures, we are seeing more extreme weather, such as storms, flooding, and wildfire. These changes are already having an impact on fungi.

CLIMATE CHANGE

The fruiting season for fleshy fungi varies slightly each year depending on the weather and species, but the fruiting time for some fungi in western Europe is changing dramatically. Before 1978, the average autumn fruiting season in southern England lasted 33 days; since 2020, it has been more than 75 days. Many root partner fungi (pp.60–61) now fruit much later in the year. This is good news for mushroom hunters, but it indicates that serious changes are happening.

A lot of fungi – especially those that decay wood and other plant material, such as the sulphur tuft fungus (*Hypholoma fasciculare*) – now fruit in the spring as well as in the autumn. This is probably because winters are a lot milder than they used to be. One consequence is that the fungus can grow and feed in winter now, so wood will likely decay more quickly, meaning carbon dioxide will be released into the atmosphere more quickly. This is not a problem if trees make wood at the same, quicker rate, but if they don't speed up then the climate may warm even more quickly than it already is.

RANGE CHANGES In a changing climate, plants follow the temperatures that suit them, moving polewards and to higher elevations in warmer climates. Fungi track the plants. Warmer temperatures also threaten the habitats of fungi that live in cold conditions, such as at the top of mountains and in high latitudes.

St George's mushroom (*Calocybe gambosa*)
While most fruiting occurs in autumn in temperate climates, some fungi have always fruited in spring, such as St George's mushroom. Its common British name comes from its approximate time of first fruiting in the UK: St George's day (23 April).

Sulphur tuft fungus (*Hypholoma fasciculare*)
This wood-decay fungus now fruits in spring as well as autumn. Whether it is the same individual that fruits twice in a year or different individuals fruiting in different seasons is not yet known.

Wood ear fungus (*Auricularia auricula-judae*)
The wood ear fungus used to appear almost exclusively on elder, but changes in the climate mean it now grows on many other tree species, especially beech.

Wood woolly-foot (*Gymnopus peronatus*)
In southern England, the leaf litter decay fungus wood woolly-foot used to be common under oak trees but is now more often found under beech. The reason why is still a mystery, but it is thought to be connected to climate change.

Devil's fingers (*Clathrus archeri*)
In the UK, this fungus was first spotted in about 1914, where it was believed to have arrived from northern Europe on imported woollen fabrics. It is now expanding into north-east Europe, especially in forests.

Slippery Jack (*Suillus luteus*)
Mycorrhizal with roots of conifer trees in its native range in Europe and Asia, this fungus has been introduced widely in southern Africa, and North and South America, in plantations of imported pine trees.

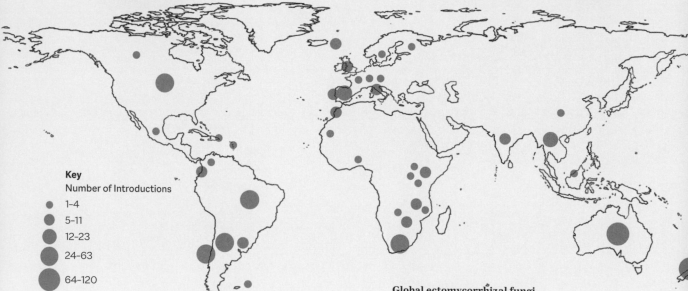

Key
Number of Introductions
1–4
5–11
12–23
24–63
64–120

Global ectomycorrhizal fungi
Ectomycorrhizal fungi are species that partner with the roots of trees (pp.60–61). Orange circles indicate countries with introduced species, the size of the circle giving an indication of the number of introduced species reported to have become established in each area.

INVADING ALLIES

Root partner fungi can sometimes help their host plants to be invasive. In the southern hemisphere, exotic pine trees from plantations are spreading into the wider landscape. They are helped by *Suillus*, *Rhizopogon*, and *Thelephora* species, which arrived on the imported trees. They quickly partner with seedlings and young trees, and co-invade with them.

Humans have inadvertently moved organisms from place to place on a small scale for millennia. But in recent centuries, international travel and trade has seen fungi move on a much larger scale.

INVASIVE SPECIES

Species newly arriving in an area often do little or no harm, but some overpopulate their new habitat. These are invasive species, which can decrease biodiversity, disrupt partnerships that have evolved over millions of years, and change the cycling of carbon and nitrogen. Invading pathogens can even change entire landscapes by killing trees (pp.78–79).

VISITOR OR INVADER? The deadly poisonous death cap (*Amanita phalloides*, pp.96–97), native to Europe, is now resident on the east coast of the USA, but it is not spreading. In California on the west coast, however, it is invasive – spreading and dominating, causing other species to decline. It was taken into the San Francisco Bay Area in the 1960s and, in highly colonized areas, almost 20 per cent of tree root tips now partner with it rather than with native fungi.

INVASIVE ROTTERS With their spectacular, often colourful fruit bodies, *Clathrus* species are easy to spot. The tentacled devil's fingers (*Clathrus archeri*) is native to southern Africa, Australia, and New Zealand. But since it arrived on the coasts of North America in the 1980s, it has become invasive. Its fruit bodies emerge from a spherical structure called an "egg", which is attached to the hidden mycelium by a cord (pp.118–119). The "egg" ruptures, and four to eight "arms" – which grow up to 10 cm (4 in) long – emerge in a starfish-like shape. Like other stinkhorns, its rotting flesh smell attracts flies to feed on the sticky upper surface where the spores are produced. Spores stick to the bodies of the flies and are spread further.

Due to land-use change, we are losing fungal habitats at rapid speed – and those that remain are being polluted or altered in other ways. All of which means that fungal biodiversity is under considerable threat.

HABITAT LOSS

Urbanization and road building erode our green spaces. Ancient forests are felled for their timber and cleared for agriculture. In the woodlands that remain, the floor is often scoured of wood for fuel or removed in the mistaken belief that "tidying up" is a good thing. Meanwhile, wetlands (pp.148–149) are drained for agriculture. Single-species, fast-growing crops are planted, and fertilizers are added to provide food for Earth's huge and expanding human population. Fossil fuels are burnt, not only causing climate change, but polluting the atmosphere with nitrogen and other noxious chemicals. Tourism tramples, compacts, and erodes soil – all of which devastates fungal habitats.

THREATENED HABITAT, THREATENED FUNGI When a habitat is lost, the fungi that depend on that habitat are lost too. The reduction in flower-rich hay meadows around the world, for example, due to fertilizer use and planting fast-growing grasses, has caused the loss of many waxcaps (*Hygrocybe* species, pp.140–141), as well as the less eye-catching club-shaped earth tongues (*Geoglossum* species, p.141).

Similarly, forests of all types, and individual ancient trees (pp.130–131), are being lost at an alarming rate. For example, the now-threatened boreal conifer forests have a history of continuous forest cover, which made them home to many rare species of fungi.

A. The globally rare ascomycete hazel gloves (*Hypocreopsis rhododendri*) looks and feels a bit like fawn to orange-brown gloves, and is a mycoparasite (pp.102–103) of glue fungus (*Hymemochaete corrugata*). It is found almost exclusively in Atlantic rainforest, a temperate rainforest that has high rainfall, mists, and lush vegetation. It is a threatened habitat – along with some of the specialist fungi it contains.

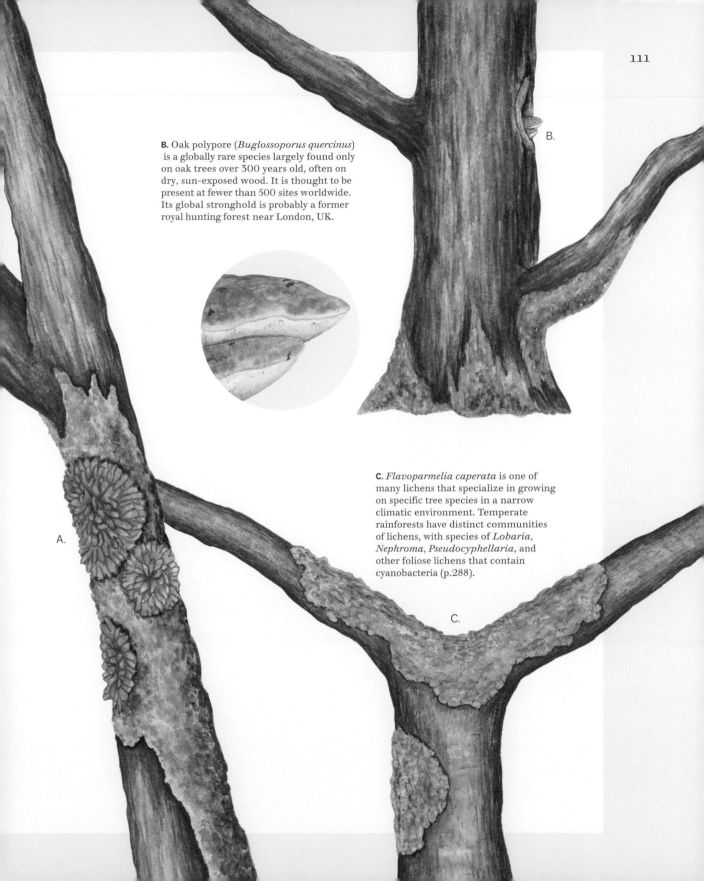

B. Oak polypore (*Buglossoporus quercinus*) is a globally rare species largely found only on oak trees over 300 years old, often on dry, sun-exposed wood. It is thought to be present at fewer than 500 sites worldwide. Its global stronghold is probably a former royal hunting forest near London, UK.

B.

C. *Flavoparmelia caperata* is one of many lichens that specialize in growing on specific tree species in a narrow climatic environment. Temperate rainforests have distinct communities of lichens, with species of *Lobaria*, *Nephroma*, *Pseudocyphellaria*, and other foliose lichens that contain cyanobacteria (p.288).

A.

C.

Dung rotters

The nail fungus (*Poronia punctata*) has nothing to do with our finger or toe nails. It is an ascomycete that grows on horse and donkey dung and looks like the head of a tiny carpenter's nail (0.5–1.5cm, ¼–⅜ in across). It was a common sight in the 19th century but, now we use motorized vehicles, its dung habitat has gone. It is on the Red List in many European countries and is now rarely seen.

Zeus olympius

They don't come much rarer than this cup fungus, which has only ever been found on Mount Olympus in Greece and in the Slavyanka Mountains in Bulgaria, on dead twigs and small branches of Bosnian pine. Its habitat is threatened by climate change and forest fires.

Bleeding tooth (*Hydnellum peckii*)

Many fungi that form root partnerships with trees on soils low in nutrients, such as boletes, amanitas, and tooth fungi, have dramatically declined as a result of fertilizer additions and air-borne nitrogen pollution from burning fossil fuels.

Fungi face extinction on both a local and, for some, a worldwide scale. But it isn't all bad news. Thanks to action already taken, some fungi are making a comeback.

THREATENED SPECIES

After much lobbying, the International Union for the Conservation of Nature (IUCN) now recognizes fungi on an equal footing with flora and fauna, but there's a lot of catching up to do. The IUCN has developed a system of Red Lists, which categorize species based on the extent to which they are threatened – from least concern all the way to extinct or regionally extinct. In most countries, and on a worldwide scale, the Red List status is now being evaluated for fungal species to help prioritize protecting those that are most threatened.

We can also protect fungi wherever we live by not collecting or damaging fungal fruit bodies, not disturbing their habitats – such as leaving dead wood on the forest floor – and avoiding the use of pesticides and inorganic fertilizers.

SUCCESS STORY The noble polypore (*Bridgeoporus nobilissimus*) is a giant of the fungal world. It grows on noble fir in the Pacific Northwest of the USA and produces huge bracket fruit bodies of up to 130 kg (287 lb). In 1995, it was found in only 13 sites and was listed as endangered. Every known tree on which the fungus appeared was then protected, and by 2006 it was found at 103 sites.

Honey fungus (*Armillaria*)

CHAPTER V

WALK

WHEREVER YOU LIVE ON THIS PLANET, species of fungi are all around you. Take a walk and explore these habitats and the fungi that live there. Discover how to spot signs of mycelium in woods and grassland, and search for fungi in the most unexpected places – from caves and sand dunes to water and permafrost.

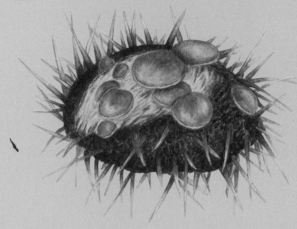

**Hairy nuts disco
(*Lanzia echinophila*)**
This little ascomycete, 1–2mm across,
can be found on the cases of sweet
chestnut and also on the cupules of
acorns. The fruit bodies are initially
orange and the upper fertile layer
(hymenium) flattens and turns
brown as it matures.

**Dark purple earth tongue
(*Geoglossum atropurpureum*)**
This little black earth tongue, up
to 6cm (2⅜ in) tall, grows in ancient
grasslands and in sand dunes. It is
difficult to distinguish from its
relatives without the aid of
a microscope.

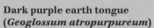

Arrhenia salina
This agaric mushroom is found at the
extreme poles, mostly north of the
Arctic Circle, and it has also been
recorded in Antarctica.

In nature, we are surrounded by fungi – though often they are hidden, becoming visible when their mushroom fruit bodies emerge, sometimes overnight. But there are some fungal fruit bodies around all the time, not just in autumn and spring.

HUNTING FOR FUNGI

This chapter will show that fungi are all around us when we explore the natural environment. Spores are in the air we breathe, and fungi exist as endophytes in plants (pp.68–69). They are in the soil under our feet, whether we are walking in grasslands, forests and woodlands, sand dunes, or even the Arctic and Antarctic. Water on boggy ground, lakes and streams, seas and oceans are also home to fungi.

On tree trunks, you can find large, long-lived brackets; on branches on the forest floor there are crusts and skins, hard warty or lumpy tissues, and rubbery and jelly-like fungi. Cast your eyes up and, perhaps with binoculars, you'll see these fungi on dead attached branches too. If you get on your hands and knees and roll over small logs, you'll see crusty fruit bodies and often mycelial cords extending into soil. Careful fingertip excavation can show the extent of these – some only tens of centimetres, others over many square metres. Look closely at leaves, cones, and other fruits, and you might uncover minute, delicate fruit bodies on thin black stalks, such as the horsehair parachute fungus (*Gymnopus androsaceus*). Even herbivore dung or a bonfire site offer fascinating fungal fruit bodies.

Other fungal signs are even less conspicuous. Blackened crusts on wood may be protective fungal coatings. Bleached patches on fallen leaves, or wood that is brown and crumbly, or stringy, or stained show that fungi are present, doing their work as decomposers. In fields, rings of lush grass or of bare earth are surface signs of mycelium below ground.

Fruit bodies are the most obvious indication of fungi, though most are only seen intermittently. If you know what to look for, you can find many other signs of fungal activity in the forest.

SIGNS OF FOREST FUNGI

A. Brown, beard-like stubble emerging from patches of sawdust or fragmented wood indicates the presence of inkcap fungi (*Coprinellus*) within. Sometimes the inkcap fruit bodies appear at the same time.

BLACK COATINGS If you stumble across tree stumps, fallen branches, or trunks that are blackened, you might think they have been charred. But these are not usually the remnants of a fire. Instead, the wood is covered by a thin, black layer of tightly woven hyphae impregnated with melanin (a dark, protective pigment) and water-repellent proteins called hydrophobins. Patches of the black tissue can sometimes be pulled off as thin plates.

This tissue is produced by some fungi that feed on wood, particularly honey fungus (*Armillaria*) and the candlesnuff fungus (*Xylaria hypoxylon*), which has white, claw-like fruit bodies (p.48). It keeps other fungi out and maintains suitable moisture conditions for the fungus: wet wood for honey fungus and dry for candlesnuff.

BOOTLACES, CORDS, AND FURRY FEATURES Remove bark from a standing dead or fallen trunk, and you might discover black, root-like bootlaces (rhizomorphs; see right). Often reddish-brown when young, and 1–2 mm in diameter, these string-like structures spread as a network through the forest floor.

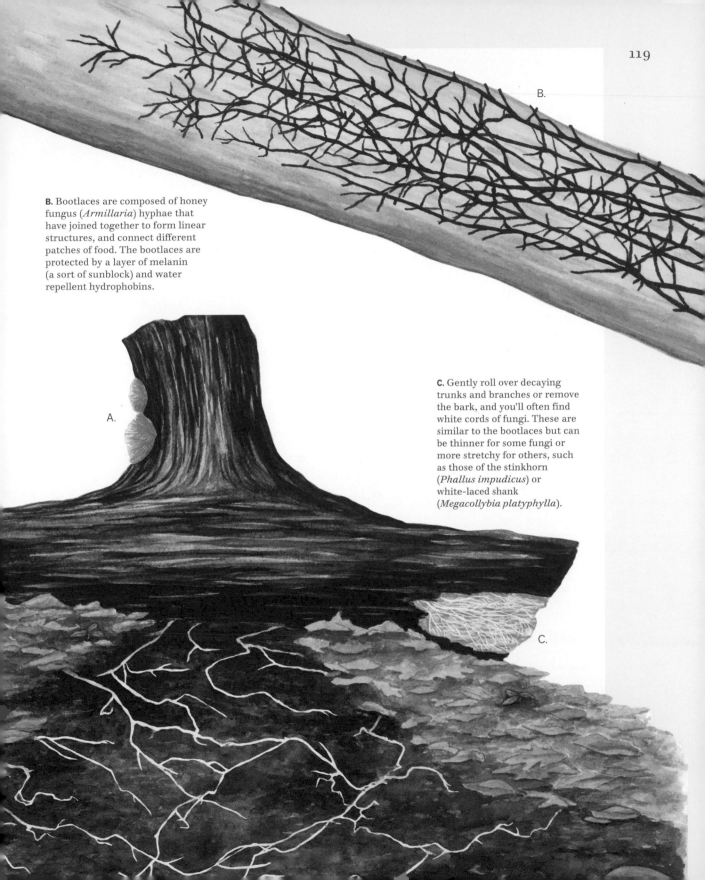

B. Bootlaces are composed of honey fungus (*Armillaria*) hyphae that have joined together to form linear structures, and connect different patches of food. The bootlaces are protected by a layer of melanin (a sort of sunblock) and water repellent hydrophobins.

C. Gently roll over decaying trunks and branches or remove the bark, and you'll often find white cords of fungi. These are similar to the bootlaces but can be thinner for some fungi or more stretchy for others, such as those of the stinkhorn (*Phallus impudicus*) or white-laced shank (*Megacollybia platyphylla*).

On the forest floor, patches of leaves and sawn
or broken wood often reveal the hidden presence
of fungi. Tell-tale signs are lines and patches
of stain and decay.

SIGNS IN ROTTING
WOOD AND LEAVES

ROTS Light patches on dead leaves and bleached wood on the forest floor, which often feels fibrous, indicate white rot. Here, a fungus – typically a basidiomycete – is breaking down all of the different types of chemical components in the wood, including the extremely complex lignin molecules, which are key structural materials. Only a narrow range of basidiomycetes are able to break down lignins. For brown rot (see below right) the decay rate – that is, how quickly weight is lost – is sometimes slower than in white-rotted wood, but it loses strength rapidly, which is a particular problem in wooden infrastructure such as buildings and transmission poles. Like white rot, only a narrow range of mostly basidiomycetes can cause this type of decay.

DECORATIVE DECAY Other signs can only be seen inside the wood, so look carefully at the cut ends of logs stacked in piles. Some sac fungi, such as *Ceratocystis* and *Ophiostoma* species (p.82), stain the wood, often a dark black or blue colour, which reduces the value of timber due to its changed appearance.

Other fungi have more appealing effects (see above right). Likewise, when the decaying wood of many broadleaf trees is cut open, it sometimes reveals intricate lines from black to brown to orange. These indicate the sites of fungal battles that have resulted in a draw or stalemate (pp.102–103) and demarcate the territories of individual fungal mycelia – much like the way in which we might build walls around our properties or towns in years gone by. Beautiful objects can be made from the wood, showcasing the striking patterns.

Nature's jewellery
A few fungi, such as the turquoise-coloured green elf cup (*Chlorociboria aeruginascens*) turn wood a green colour thanks to the chemical dyes produced by their mycelium. The wood is hardly decayed, and its colour is prized for making decorative items. It is easy to spot both on the forest floor and in antique boxes, such as Tunbridge Ware.

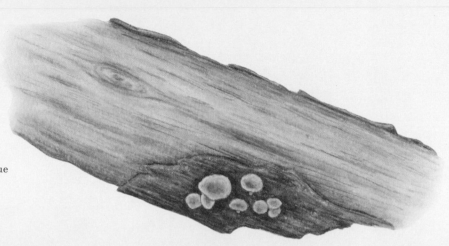

Brown rot
Wood that is brown and crumbly with cubic cracking, often seen in decaying conifer and oak wood, is being rotted by fungi that break down the main chemical components except lignin. This is brown rot.

Tar spot
The biotrophic fungus *Rhytisma
acerinum* is most commonly associated
with the black marks of tar spot on
sycamores (above), but it also causes
patches of green on ageing sycamore
leaves (right).

Autumn leaves are often yellow and amber but, look closely at those on the forest floor, and you might see patches of green. These so-called green islands are evidence of fungi.

GREEN ISLANDS

The green islands that you can spot on autumn leaves are sometimes caused by mosaic viruses, or by the larvae of leaf miner insects such as *Phyllonorycter blancardella* moth on apple leaves. However, most of these green islands are produced by fungi, particularly those that feed as biotrophs (on living tissues) for at least part of their lifecycle. These fungi include rust diseases such as *Puccinia graminis*, which infects wheat, and powdery mildews (pp.72–73) such as *Blumeria graminis*, which infects barley. Some of the most striking green islands are those produced by the ascomycete *Rhytisma acerinum* (see left).

HOW GREEN ISLANDS FORM Plants produce hormones that act as chemical messengers to control processes such as plant growth and development and how plants respond to environmental stresses. Leafy green tissue signifies active photosynthesis, stimulated by hormones called cytokinins. As photosynthesis is occurring, senescence (ageing) is supressed. Once a plant approaches dormancy, cytokinin levels drop, reducing photosynthesis and triggering nutrients to move from the leaves to other parts of the plant. The leaves yellow and eventually fall.

Green islands are sites where photosynthesis is still occurring, even when a leaf has fallen: the fungi responsible are keeping plant cells alive so they can feed off living plant tissue until the leaf dries out or its resources are depleted. Many of the fungi capable of producing green islands can make cytokinins, as well as enzymes that influence the action of cytokinins; it's not yet known if these are produced in the leaf by the invading fungus or if the fungus forces the plant to produce them.

ARMILLARIA IN HISTORY

One of the earliest records of fungal bioluminescence comes
from the Greek philosopher Aristotle, who wrote about a
"glowing foxfire", presumably referring to wood colonized
by the honey fungus *Armillaria* species (above). It is also said
that timbers infected with *Armillaria* species provided natural
torchlight for British soldiers in the trenches during the First
World War. The same fungus hampered the Allied war effort in
the Second World War: infected timber is alleged to have guided
enemy aircraft towards the London timber yards.

Many marine organisms, fireflies, fish, algae, and bacteria emit a greenish light and therefore glow in the dark. This captivating ability, called bioluminescence, is shared by some species in kingdom fungi too.

FUNGI THAT GLOW

Not all fungi glow in the dark, but all of those that do are mushroom-forming, white-rot wood-decay fungi (pp.120–121). The mycelium or fruit bodies, or both, can glow, depending on the species. The bonnet fungi (*Mycena*), of which more than 70 species are bioluminescent, produce light from their fruit bodies, whereas the European Jack-o-lantern (*Omphalotus olearius*) emits light throughout the whole body of the fungus, including both the mycelium and its mushrooms.

WHY DO SOME FUNGI GLOW? Recent research suggests that bioluminescence in fungi happens when they recycle a valuable "mopping up" agent, the antioxidant hispidin. When white-rot fungi decompose wood, they produce enzymes to break bonds in the polymer lignin, a complex molecule that is a major component of wood. In doing so, they inadvertently produce highly reactive chemicals called free radicals, which are very toxic to fungal cells. To avoid "death by decomposition", the white-rot fungi have two choices. The first option is to produce lots of hispidin, which makes the free radicals less reactive so they don't damage cells, as is seen in the white-rot bracket fungus *Inonotus hispidus* (from where hispidin was first isolated). White-rot, mushroom-forming fungi do not contain high levels of hispidin, so they use a second option and recycle it in the fungal cells. This is where bioluminescence comes into play. The hispidin recycling process forms a light-emitting chemical, giving the fungus a captivating glow.

In forests, we are surrounded by fungi. They are in the trunks and branches of standing trees above us, in fallen wood, decaying leaves, and twigs beneath our feet, and attached to tree roots.

ANCIENT OR MANAGED FORESTS

Fungal species in forests vary depending on the region and what species of trees and other vegetation are in the forest. Another important factor is the age of the forest and whether it is managed.

ANCIENT FORESTS In ancient, unmanaged forests, the floor is littered with large trunks that can take decades, if not centuries, to rot. These might just display the decay caused by fungi that were in the standing tree (pp.130–133) or they might be crumbly remains more like soil than wood, a shadow of a former presence. These forests host a huge diversity of fungi, only found where there are large, long-lasting logs. This continuity is vital as wood-decay fungi need a long-term supply of suitable dead wood. Ancient beech forests contain rarities including, on fallen trunks, the tiny green pimple-dotted jelly (*Hypocrea gelatinosa*), the late fall or resinous polypore (*Ischnoderma resinosum*), and *Pluteus hispidulus* on decayed wood.

MANAGED FORESTS Most woodland has been managed for 2,000 years, for wood for buildings, furniture, or fires, or in the mistaken belief that tidying up fallen wood is a good thing. But a careful look at stacks of logs and tree stumps reveals a treasure trove of fungi (see right).

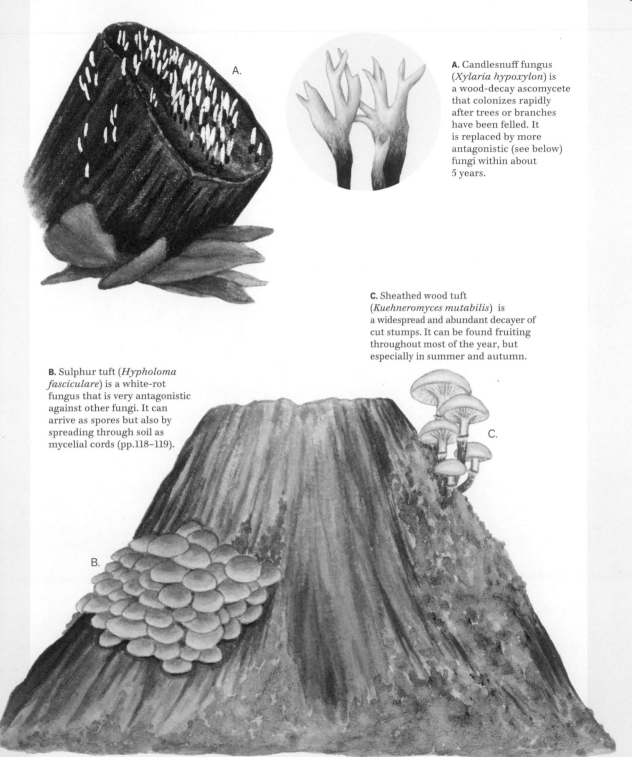

A. Candlesnuff fungus (*Xylaria hypoxylon*) is a wood-decay ascomycete that colonizes rapidly after trees or branches have been felled. It is replaced by more antagonistic (see below) fungi within about 5 years.

C. Sheathed wood tuft (*Kuehneromyces mutabilis*) is a widespread and abundant decayer of cut stumps. It can be found fruiting throughout most of the year, but especially in summer and autumn.

B. Sulphur tuft (*Hypholoma fasciculare*) is a white-rot fungus that is very antagonistic against other fungi. It can arrive as spores but also by spreading through soil as mycelial cords (pp.118–119).

Marasmius tageticolor
Although its fruit body has the familiar umbrella
shape of its temperate counterparts, this fungus
has striking red-and-buff white stripes.

They might cover less than 6 per cent of the planet's land surface, but tropical rainforests are host to an almost unimaginable variety of plant, animal, and fungal life.

TROPICAL RAINFOREST

Situated between the tropics of Cancer and Capricorn, tropical rainforests have shrunk to less than half their original size, but these vast areas are still much the same as they were 70 million years ago. Throughout the year, rainforests are warm, wet, humid, and cast in deep shade.

BIODIVERSITY The plant and animal diversity is astounding: the Amazon rainforest contains over 40,000 plant species, including almost 12,000 tree species, 1,300 bird species, and thousands of invertebrate species. Every plant and animal species is likely to have at least one fungus associated with it, so the fungal diversity is huge. Most tropical fungi still don't have names; one in five of the species found are new to science. While some are recognizable as relatives of familiar temperate species, others are dramatically different in form or shape and patterning.

As in temperate forests, high-rise fungal inhabitants are at work decaying twigs and branches, while endophytes (pp.68–69), lichens, and pathogens are on or in the trees and shrubs. To avoid the intense competition for dead plant material on the forest floor, some decayers form mycelial nets to trap falling plant debris. The floor fungi are mostly decomposers. Some colonize fresh material, and others only well-decayed material, and many are specific to a plant species or type of dead material. This leads to a mosaic of fungal species through the forest; the species in a single square metre would be very different to those in a similar-sized patch just 100 m (330 ft) away. Though fruit bodies can be found all year, there are periods of more intense fruiting that are much longer than those in temperate forests.

Forests are living history. While some trees have relatively short lives, others – such as oak, yew, and red cedar – can live for a thousand years or more. In all stages of a tree's later life, fungi play a crucial role.

AGEING TREES

MATURE, VETERAN, AND ANCIENT TREES Like us, trees change as they mature. Small twigs and branches die as the branches above limit their sunlight. A veteran tree has a broader trunk, a shorter and squatter shape, and a reduced canopy. In the ancient phase, the trunk hollows.

BARK Cankers are areas of dead, often sunken, bark, mostly caused by pathogens that kill the living parts of bark. A few can then grow deeper, affecting the underlying wood. But most wood-decay fungi that cause cankers grow outwards from colonized inner wood into the bark. The ascomycete *Eutypa spinosa* forms long strip cankers in this way, which are sometimes mistaken for lightning strikes.

HOLLOWING The centres of living trees, called heartwood, are dead, so wood-decay fungi living there mostly do not harm the tree, but release nutrients that it can reuse. Heartwood of some trees, such as oak, contains tannins and other chemicals that deter many fungi and invertebrates, so only a few fungi have evolved to grow in this type of wood. Others, such as beech, do not contain toxic chemicals and are colonized by a range of fungi. The decay forms hollows, offering habitats for many birds, mammals, and invertebrates.

WOODY ROOTS Fruit bodies at the base of a tree show that large woody roots may be rotting; decay may extend into the trunk.

A. The porcelain fungus (*Mucidula mucida*) is a basidiomycete fungus. Native to Europe, it usually appears on beech wood in late summer and autumn.

C. Red-belted polypore (*Fomitopsis pinicola*) causes brown rot of dead wood in living conifer and broadleaf trees. It is common in the northern hemisphere, but rare in the UK.

F. *Ganoderma australe* is a tough bracket fungus that survives for several years. It decays the dead heartwood of the trunk.

B. Beech woodwart (*Hypoxylon fragiforme*) is an ascomycete that decays dead branch wood. It can be spotted all year. Each is only 2–9 mm (up to ⅓ in) across, but they form in patches.

D. Beech tarcrust (*Biscogniauxia nummularia*) is an ascomycete wood-decay fungus. Its fruit bodies are thick, black crusts.

E. Chicken of the woods (*Laetiporus* species) is found worldwide. The bright yellow, fleshy brackets fade with age.

G. *Meripilus giganteus* is a bracket fungus, common on broadleaved trees such as beech and oak. It rots the woody roots.

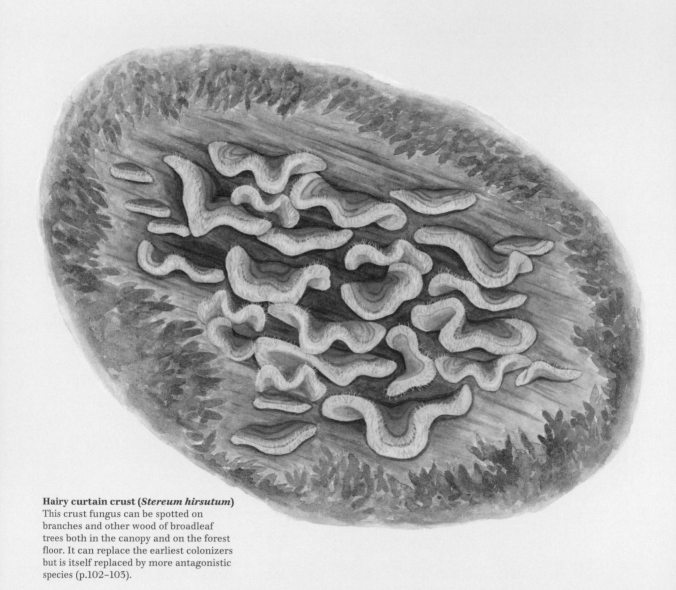

Hairy curtain crust (*Stereum hirsutum*)
This crust fungus can be spotted on
branches and other wood of broadleaf
trees both in the canopy and on the forest
floor. It can replace the earliest colonizers
but is itself replaced by more antagonistic
species (p.102–103).

From dead branches in the canopy to rotting wood on the forest floor, different communities of fungi are responsible for decaying wood at different stages.

DECAYING BRANCHES

CHANGES OVER TIME Decay begins while dying branches are still attached in the canopy, and the species of fungi in the wood change as months pass. The first fungi are adapted for the environment: some cope well with the high water content of recently living branches where they started their lives as endophytes (pp.68–69); others make masses of spores that germinate quickly so they can grow before other fungi arrive. Later, fungi arrive as spores or, sometimes, as mycelium via soil (pp.118–119). If these are better fighters (pp.102–103) than the early colonizers, they soon replace them. Specialists in well-decayed wood may then come to dominate. Although these changes are hidden inside the wood, if you return over several years you'll see evidence in the form of a succession of different fruit bodies.

SPECIALISTS Different tree species have characteristic communities that start the decay process, visible by their fruit bodies. In oak, these are crusts of *Stereum gausapatum*, *Peniophora quercina*, and the pale sometimes violet-tinged, waxy-coated branches harbouring *Vuilleminia comedens*. Beech has the erumpent warts of the ascomycetes *Biscogniauxia nummularia* and *Hypoxylon fragiforme*, or the almost translucent porcelain fungus (*Mucidula mucida*, pp.130–131). The winter fungus or velvet shank (*Flammulina velutipes*) is common on the branches and stumps of many broadleaf trees. It brightens a winter walk when its bright orange-coloured caps appear. In the canopy, these fungi may be replaced by common crust fungi such as turkey tail (*Trametes versicolor*). These crust fungi can finish the decay process, but they are often replaced by even more antagonistic fungi such as cord-forming (p.127) sulphur tuft (*Hypholoma fasciculare*) on the forest floor.

Look closer at the smaller components of forest life – small dead parts of plants, such as cones, nuts with hard shells, attached dead twigs, or even the thicker stems of some herbaceous or shrubby plants – and you'll find fungi there too.

CONES, NUTS, AND TWIGS

Some fungi specialize in feeding on small litter components and have evolved to cope with the unusual chemical composition, small size, hardness, and amount of water these plant tissues contain. Like their larger counterparts, they too play important roles as recyclers.

Ear-pick fungus or cone tooth (*Auriscalpium vulgare*) has an unusual fruit body, with a stem to the side of the cap and tooth-like projections rather than gills beneath. It is found on pine cones or their remnants on the forest floor or sometimes completely buried in soil, but it can also appear on cones of Douglas fir, and occasionally others such as spruce. It is widespread in Europe, Central and North America, and temperate Asia.

Hymenoscyphus fructigenus specializes in the decay of nuts, especially beech and hickory shells, and acorns, where it appears as clusters of tiny white cups. *Colpoma quercinum* is an ascomycete with minute, oval-shaped crusts (2–15 mm, or ½ in long). It is an endophyte (pp.68–69) in healthy oak twigs and then an early decayer of dead twigs that are still attached to the tree.

**Beechmast candlesnuff
(*Xylaria carpophila*)**
Resembling the closely related ascomycete
Xylaria hypoxylon (p.127), commonly seen
on wood, beechmast candlesnuff specializes in
feeding on the hard outer cases (cupules) that
enclose beech seeds. Visible throughout the
year, the fungus turns black in autumn and
early winter when it produces sexual spores.

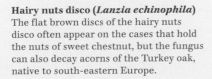

Catkin cup (*Ciboria batschiana*)
This ascomycete has round brown cups that emerge
on a stem up to 2 cm (¾ in) long from old acorns and
sweet chestnut nuts.

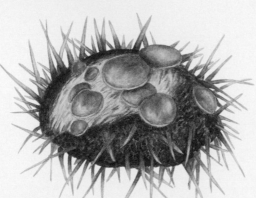

Hairy nuts disco (*Lanzia echinophila*)
The flat brown discs of the hairy nuts
disco often appear on the cases that hold
the nuts of sweet chestnut, but the fungus
can also decay acorns of the Turkey oak,
native to south-eastern Europe.

**Marasmius capillaris*
Marasmius capillaris and the very similar
Mycena polyadelpha decay leaves of beech
and oak respectively, producing their fruit
bodies in late summer to winter. Tiny
white, umbrella-shaped fruit bodies
appear on the end of long, thin stems.

Fungi can exist without forests, but forests can't exist without fungi. Integral to the forest ecosystem, fungi grow in various ways.

WOODLAND RINGS

Mycelium of some forest-floor fungi grows in individual dead plant parts (pp.134–135), but it more often colonizes large patches, sticking leaves together in clumps. The fruit bodies appear in patches or as isolated individuals, small groups or troops, or even as distinct rings.

FREE RINGS Some decomposer fungal species form rings of fruit bodies of a single individual fungus. The mycelium starts from a single point of origin in the centre. It first forms a small, roughly circular patch, which moves ever outwards. The mycelium in the middle dies, and it then extends as a wide band of mycelium, leaving an empty central circle as it grows outwards. As the outer margin extends, the hyphae in the inner edge die and the hyphal material is recycled for use elsewhere in the mycelium. Fruit bodies appear in the band of mycelium in autumn.

These fruit bodies indicate the presence of the hidden mycelium beneath the surface, feeding on dead leaves and other plant parts. When the fairy ring encounters a rock or a tree, the ring continues to extend on either side, but the mycelium does not join up again, so a gap is left in the ring. It is not always easy to spot a woodland fairy ring, whereas rings in grasslands often give away more clues (pp.138–139).

UNSOLVED MYSTERY
Scientists once thought that fairy rings of decay fungi move outwards because they run out of food or because the central circle becomes toxic. However, as new dead leaves drop every year, fungi cannot run out of food in woodlands. A simple experiment also suggests that the centre is not toxic: if a turf containing the full width of the band of mycelium is moved into the ring's centre, the fungi can still grow. Intriguingly, when the turves are placed facing different directions, the mycelium only grows from the side that was originally the leading edge.

Tethered rings
These tethered rings are produced by fungi, such as this *Amanita muscaria*, around trees. Typically, these grow in ectomycorrhizal partnership (pp.60–61) with the tree roots. The fruit bodies do not mark the edge of the mycelium, which may be many metres further out, taking up water and nutrients which it passes to the tree in exchange for sugars.

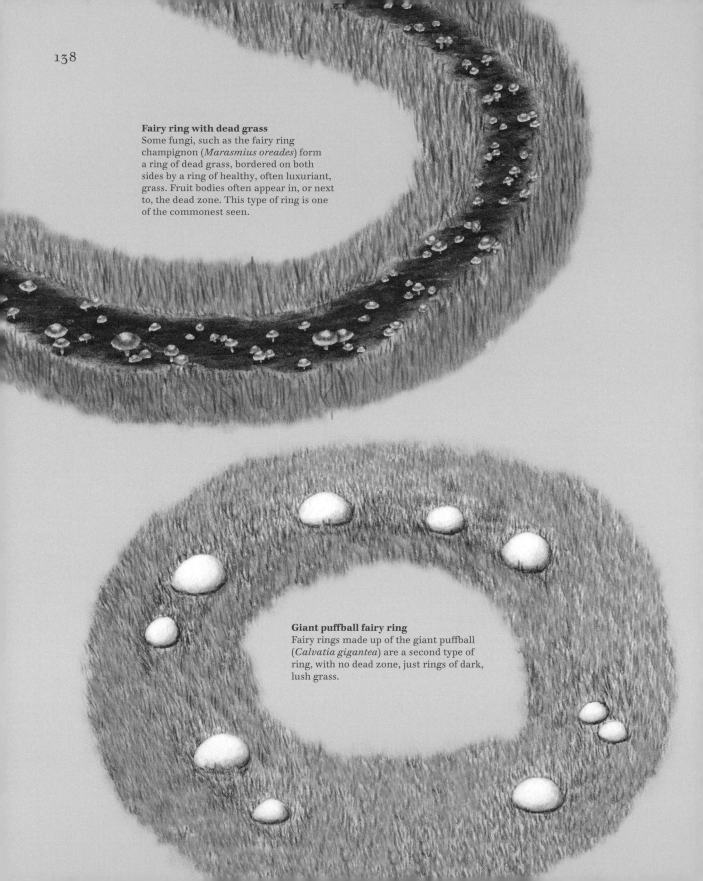

Fairy ring with dead grass
Some fungi, such as the fairy ring
champignon (*Marasmius oreades*) form
a ring of dead grass, bordered on both
sides by a ring of healthy, often luxuriant,
grass. Fruit bodies often appear in, or next
to, the dead zone. This type of ring is one
of the commonest seen.

Giant puffball fairy ring
Fairy rings made up of the giant puffball
(*Calvatia gigantea*) are a second type of
ring, with no dead zone, just rings of dark,
lush grass.

Rings of mushrooms have captivated humans' imaginations for centuries. But some fairy rings are invisible – almost. Subtle signs in grassland indicate the fungal presence beneath the surface.

GRASSLAND: FAIRY RINGS

The biggest fairy rings are hundreds of metres from one side to the other. Some are even visible on Google Earth, such as those at the Laramie Regional Airport, Wyoming, USA, around and beyond the westernmost runway.

The mycelium grows outwards from a central point of origin, and inner regions of the mycelium die as it advances outwards – hence why the fruit bodies only appear at the edge. More than 140 species form rings and they manifest in three main ways. Firstly, as dead or dying grass killed by the fungus (see top left); secondly, as lush rings of grass that have benefitted from nutrients being released into the soil by fungi breaking down dead plant material (see bottom left); and thirdly, without any visible effects on the grass, but with rings of fruit bodies appearing at appropriate times of year.

Fairy rings expand at a rate of around 4 cm (1½ in) every year, and the largest are estimated to be at least 2,000 years old. When different rings meet each other, they die where they make contact, but the rest of the ring continues to grow in arcs and complex shapes.

Plants in grasslands are very different from those in forests.
They have different root systems and little woody tissue.
As a result, the fungal specialists that grow in
grassland are different too.

ANCIENT GRASSLANDS

Grasslands cover about 20 per cent of the global land surface, but most have been managed by humans, who plough, reseed with fast-growing grass species, and add inorganic fertilizer. Very little of Europe's grassland has been left alone for more than 100 years. This more natural grassland includes large, flower-rich hay meadows, lawns, or even grass in graveyards, all of which can be a sanctuary for fungi. Fungal species in grasslands vary depending on climate and soil type – specifically, whether the land's soil is acidic or chalky (calcareous) – but more natural grassland is home to the rarest and most charismatic of grassland fungi.

CHEGD FUNGI More natural grassland is recognizable by the presence of a large number of CHEGD fungi: the more there are, the more important it is to protect the site. The CHEGD abbreviation is simply the first letter of the characteristic groups of fungi found in these grasslands:

• C – clavarioid fungi (coral- and club-shaped fungi, pp.198–199): these range from off-white to brightly coloured. Some are globally vulnerable (pp.112–113), such as the distinctive violet coral (*Clavaria zollingeri*, p.47).
• H – *Hygrocybe* (waxcaps) and relatives: with their waxy, often brightly coloured fruit bodies, these fungi are always an exciting find – whether it be the scarlet hood (*Hygrocybe coccinea*) or the rare, globally vulnerable reddish-purplish jubilee waxcap (*Gliophorus reginae*). In fact, 90 per cent of *Hygrocybe* species are on the Red List of at least one European country.

- E – *Entoloma* (pink gills): for example, the big blue pink gill (*Entoloma bloxamii*).
- G – *Geoglossum* (earth tongues) and relatives: these ascomycetes are often hard to spot, with their black, dark brown, or olive-green slightly flattened fruit bodies that only protrude about 2.5–7 cm (1–2¾ in) above the soil.
- D – *Dermoloma* (crazed caps) and relatives: these are small grey-brown mushrooms, often with cracks in the surface of the cap.

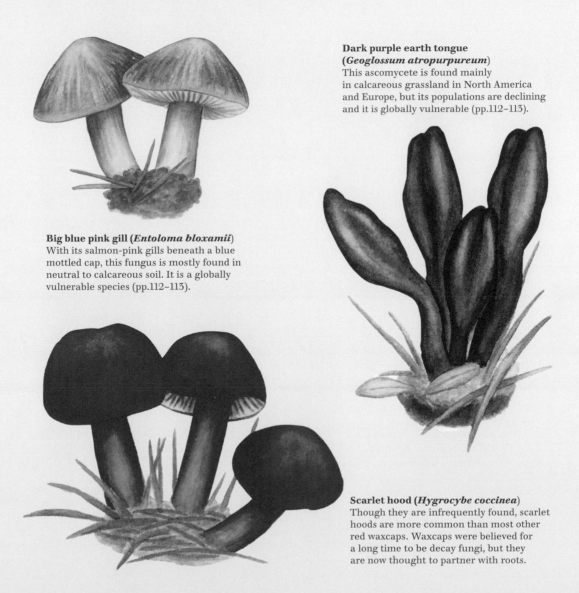

Dark purple earth tongue (*Geoglossum atropurpureum*)
This ascomycete is found mainly in calcareous grassland in North America and Europe, but its populations are declining and it is globally vulnerable (pp.112–113).

Big blue pink gill (*Entoloma bloxamii*)
With its salmon-pink gills beneath a blue mottled cap, this fungus is mostly found in neutral to calcareous soil. It is a globally vulnerable species (pp.112–113).

Scarlet hood (*Hygrocybe coccinea*)
Though they are infrequently found, scarlet hoods are more common than most other red waxcaps. Waxcaps were believed for a long time to be decay fungi, but they are now thought to partner with roots.

Every plant in the garden is associated with a fungus – from mycorrhizal fungi and endophytes supporting plant growth, to disease-causing pathogens and decomposers clearing up plant waste.

GARDENS AND LAWNS

Thousands of different fungi can be found in gardens and parks, allotments and flower beds, even in pots on the patio or window boxes, and in compost heaps, log piles, and lawns. These fungi often remain hidden from view, yet they form an integral part of the garden's tapestry.

LAWNS AND TURF GRASS Dead organic matter in lawns and grass is consumed and recycled by fungi, some forming fairy rings (pp.138–139). Others are pathogens on lawns and turf grass. Fusarium patch is caused by *Microdochium nivale* and is often spotted after snow thaw – hence its alternative name, snow mould. The fungus causes lawn grass to yellow and then dry out and turn brown, often in large patches. Occasionally, its pinkish mycelium is seen, particularly around the edges of the patch. Another fungus that causes yellow patches on grass is red thread disease, caused by *Laetisaria fuciformis*. Two stages of this fungus's life cycle can be seen. First, the fungus produces bright red structures (sclerotia) resembling fine needles that protrude from the blades of grass. These fungal survival structures (p.289) can survive for long periods in the soil. Second, when they germinate, hyphae quickly invade fresh grass through small openings on the blade surface called stomata.

UNINVITED GREENHOUSE GUESTS Fungal fruit bodies often found in the tropics and subtropics can appear in greenhouse compost. For example, the deadly *Conocybe rugosa* (p.96) can sometimes be found on compost across mainland Europe, North America, and Africa.

Honey fungus (*Armillaria*)
There are several species of honey
fungus that often emerge on lawns.
Some are killers of tree roots, shrubs,
and even non-woody plants. Others are
opportunists, colonizing roots of trees
that are weak or have recently died.

**Red thread fungus
(*Laetisaria fuciformis*)**
This fungus is sometimes
confused with snow mould
because the fluffy pink
mycelium of red thread fungus
can often be seen at the base
of each blade of infected grass.

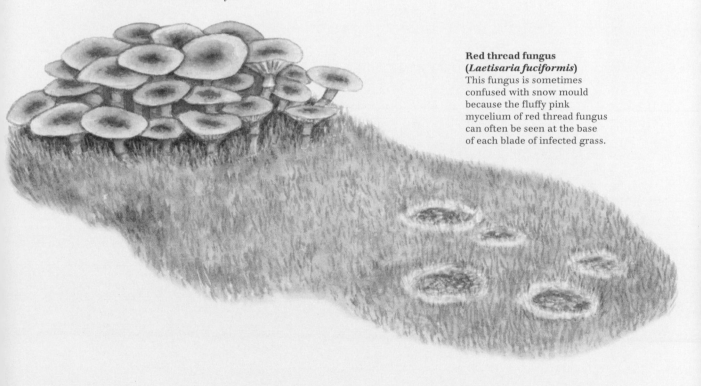

Egghead mottlegill (*Panaeolus semiovatus*)
The dull buff-coloured fruit bodies of this
coprophilous (dung-loving) fungus rots
herbivore dung in Europe and North America.
It does not appear until at least 10 days after
dung has been deposited.

Next time you're watching where you're stepping in a field peppered with animal dung, keep an eye out for fungi. Dung is home to some specialist species.

HERBIVORE DUNG

Herbivore dung is nutrient-rich, alkaline, and moist – a good home for specialist fungi. The composition of dung depends on the animal species that deposited it, how long it is on the ground, and the environmental conditions – all of which affect which fungi grow on it. When dung lands, it already contains spores of many dung-decay fungi. The spores germinate, and hyphae colonize the dung. Fungi feed and grow at different speeds. Some only break down and use simple chemical molecules as food, while others can also use more complex molecules.

FRUITING SEQUENCE When the fungi have enough nutrients, they make fruit bodies. Those that use simple chemical compounds and make small fruit bodies produce them in a few days. Those that use more complex compounds and make larger fruit bodies grow for several weeks before the fruit bodies appear. Minute fruit bodies of species of Mucoromycota *Mucor*, *Pilaria*, and *Pilobolus* appear 1–3 days after dung has been deposited, disappearing after 7 days. The fruit bodies of slightly larger (a few millimetres across) ascomycetes, such as *Ascobolus* and *Cheilymenia*, appear after 5–6 days, with species such as *Podospora* and *Sordaria* by 9–10 days. After 10 days or more, the dung-specialist basidiomycetes with their larger fruit bodies appear, including the midden inkcap (*Coprinus sterquilinus*) with its shaggy ovoid cap, and the dung roundhead (*Protostropharia semiglobata*), which is smooth and shiny when dry but sticky when wet.

LIFE CYCLE OF DUNG FUNGI Spores are usually consumed when the herbivore eats vegetation, but most herbivores don't like to eat near their own dung. To get around this, the hat thrower (*Pilobolus*, p.41) shoots its package of spores over 1 m (3 ft) away from its delicate fruiting structures.

Like some plants, certain fungi thrive after fires. These are called phoenicoid (meaning arising from the ashes) or pyrophilous (meaning burn-loving) fungi.

FIRE-LOVING FUNGI

Fires burn 570 million hectares (1,400 million acres) of land every year, either through prescribed burning or wildfire. In ecosystems that rarely experience wildfires, fungal communities are destroyed and take many years to recover. But where fires occur regularly, fungi have evolved to cope. There are different types of phoenicoid fungi. Some are seen exclusively on burnt ground, such as the tiny olive-brown discs of the ascomycete *Ascobolus carbonarius,* which blend in with scorched earth. Others – including the stalked bonfire cup (*Geopyxis carbonaria*), with its small, brownish goblet-shaped cups with a whitish rim and often in prolific numbers – most often appear on burnt ground. Different species of phoenicoid fungi cope with heat in different ways. *Cenococcum geophilum*, for example, survives in mycorrhizal roots or as mycelium deeper in soil.

RESPOND Some fungi don't just survive after fire but thrive. Certain species are triggered to grow or fruit by intense heat or by chemical changes in the environment caused by heat. *Neurospora crassa* is easily spotted after vegetation has burned. It coats wood in masses of orange patches of mycelia and asexual spores. Its ascospores (p.55), with their black walls and thick ribs, are stimulated to germinate by temperatures of 50–70°C (122–158°F). Bonfire scalycap (*Pholiota carbonaria*) can live as an endophyte (pp.68–69) in mosses, but only makes fruit bodies when the mosses have burned.

WHEN AND WHERE TO SEARCH

Even small bonfire sites yield interesting finds, but not until 7 or so weeks afterwards. *Anthracobia* and *Pyronema* species appear early but are gone after 12 to 18 months. Some, such as *Ascobolus carbonarius, Peziza* species, and *Pholiota carbonaria,* emerge after 10–15 weeks but are rarely seen after 2 years. Finally, *Plicaria endocarpoides* doesn't appear before 20–50 weeks, though it is sometimes still seen 3 or 4 years after a bonfire.

A. *Plicaria endocarpoides* is a common European species producing dark brown cups, up to 6 cm (2⅓ in) across, found on burnt ground from spring to autumn. It does not appear until 20–50 weeks after a fire but may persist for up to 3 or 4 years.

B. Charcoal eyelash (*Anthracobia melaloma*) has small (less than 0.5 cm or ¼ in across) orange, cup-shaped fruit bodies that are found almost exclusively on burnt ground, including bonfire sites and on rocks following volcanic eruptions. It has heat resistant spores.

C. Some ascomycete morels (*Morchella*), with their honey-combed cups on stems, arise phoenix-like from the ashes of burnt conifer forests. Several can survive wildfires, such as *Morchella tomentosa* in western North America. This is thanks to their buried survival structures called sclerotia, which are a mass of spherical or slightly flattened fungal tissue with a thick outer layer.

**Midnight disco
(*Pachyella violaceonigra*)**
Found on submerged and wet wood,
the rare midnight disco ascomycete
fungus can be recognized by its
large (several centimetres across),
gelatinous, violet-coloured cups.

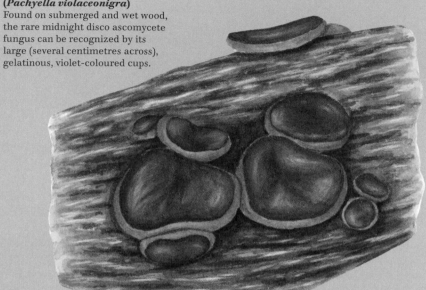

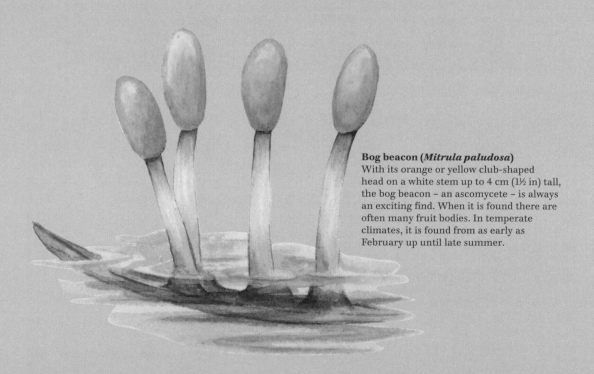

Bog beacon (*Mitrula paludosa*)
With its orange or yellow club-shaped
head on a white stem up to 4 cm (1½ in) tall,
the bog beacon – an ascomycete – is always
an exciting find. When it is found there are
often many fruit bodies. In temperate
climates, it is found from as early as
February up until late summer.

*Unlike most terrestrial habitats, marshes
are – by definition – very wet. Here, you'll find fungi
that are able to cope with these particular conditions.*

MARSHY HABITATS

Marshy habitats are declining. In the last 300 years, 87 per cent of the world's wetlands have been drained for agriculture, housing, and industry. Some of the fungi from these areas are now rare. The roots of aquatic plants harbour a diverse range of fungi. Most form arbuscular mycorrhizas (pp.60–61) or are endophytes (pp.68–69), which produce spores but not macroscopic fruit bodies. The roots of alder, poplar, and willow (the main trees along waterways in many temperate areas) form ectomycorrhizas (pp.60–61) and arbuscular mycorrhizas, so fruit bodies are sometimes seen. Still, most of the fruit bodies in wet sites belong to decay fungi.

DECAY FUNGI IN MARSHLAND The matchstick-like ascomycete bog beacon (*Mitrula paludosa*) is widespread but seen infrequently. Emerging from the dark, decaying remains of plants, algae, and mosses, the bog beacon's fruit bodies raise their heads above clean, often slow-moving water in a striking display.

Marsh honey fungus (*Desarmillaria ectypa*) grows among rotting *Sphagnum* mosses, reeds, and sedges in bogs and fens in northern Europe and Asia. Its fruit bodies are a rare find. In fact, it is on the Red List of 11 European countries. However, when it was searched for as part of the "Lost and Found Fungi" project by the UK's Royal Botanic Gardens, Kew, it was found more often than predicted.

Most fungi that rot submerged wood are ascomycetes. Many are flask fungi with small fruit bodies, but there are some larger cup fungi (pp.48–49) such as the midnight disco (*Pachyella violaceonigra*).

A perhaps unexpected place to find fungi
is in freshwater – from lakes and rivers to ponds,
puddles and even in water-filled hollows in trees.

FRESHWATER FUNGI

An average stream can contain as many as 30,000 freshwater fungal spores per litre (2 pints) of water, and the roles fungi play as decomposers in freshwater ecosystems is pivotal: rivers and streams would soon clog up if fungi didn't break down dead leaf matter in the water. Scientists have discovered over 3,000 species of freshwater fungi, including aquatic ascomycetes found on submerged wood and chytrids that produce motile spores (pp.14–15). Probably the most well-studied group are freshwater hyphomycetes, fungi that, in general, have no known sexual stage and are often referred to as Ingoldian fungi in honour of the pioneering mycologist, Cecil Terence Ingold, who first described them in detail.

Although most aquatic hyphomycetes commonly associate with freshwater, some have been found in tree holes, roots, and the canopy as well as on the forest floor. Others are endophytes (pp.68–69) living within submerged roots, for example *Campylospora parvula* and *Filosporella fistucella*, which are found in submerged aquatic roots of some alder species.

FUNGAL ADAPTATIONS Many freshwater ascomycetes have adaptions that help spores attach to woody material. Fruit bodies (pp.48–49) are often partially or fully seated within the woody substrate, and some have sacs (asci) (pp.176–177) that deliquesce (break down), aiding spore release.

Spores of Ingoldian fungi are also well adapted to aquatic environments, having relatively large, uniquely shaped spores, often reaching lengths of 100 μm (one tenth of a millimetre). Any fast-flowing river or stream will harbour spores of these fungi. The best way to see them is to use a plastic container to collect the foam that forms as water passes over stones. Within the foam are many wonderfully shaped fungal spores, trapped by the bubbles, which you can see through a light microscope.

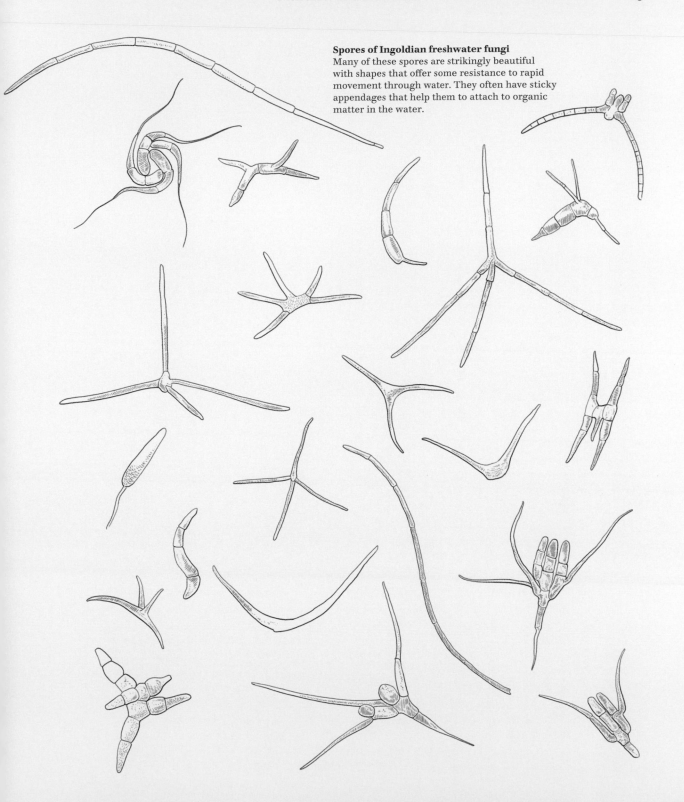

Spores of Ingoldian freshwater fungi
Many of these spores are strikingly beautiful with shapes that offer some resistance to rapid movement through water. They often have sticky appendages that help them to attach to organic matter in the water.

*Our planet's seas and oceans are teeming with fungal life
but, unlike on land, they show few visible signs – unless you
use a microscope, extract their DNA, or can culture them in
the laboratory. Exploring marine fungi is not
so much a walk as a scuba dive.*

SEAS AND OCEANS

Over 1,000 species of marine fungi are known, but scientists estimate that there
are 10,000 more still to be discovered. Half of the fungi in water samples are
species that are new to science. Some have been washed out from land, but many
are marine specialists. Some are parasites, some are decay fungi, and others
partner with seaweeds.

Microscopic chytrid fungi (pp.14–15), whose spores can swim, can parasitize
diatoms – microscopic algae less than one-fifth of a millimetre long. Fungi can
also colonize leaves or wood from mangroves, which line a quarter of the world's
coasts. The fungi can't usually be seen, but if the mangrove litter is put into a
lidded dish surrounded by moist tissue, tiny cup-shaped fruit bodies – one-tenth
of a millimetre high – often emerge.

KILLERS AND PARTNERS The most common fungi found on seaweeds are
pathogens, but a few are essential partners. *Mycophycias ascophylli*, for example,
grows in the brown alga called rockweed, which grows on shores in the intertidal
zone (the area exposed during low tide and underwater during high tide). The
fungus protects young rockweeds from drying out when the tide is low. Coral
reefs are the most biologically diverse ecosystems in the marine environment, but
they are under threat. Not only do they suffer damage from ocean acidification
and warming, but fungi such as *Aspergillus sydowii* (right) cause bleaching.

AMAZING ADAPTATIONS Marine fungi have adapted to cope with the challenges of the ocean environment. Glycerol, which is formed in fungal cells, keeps the internal saltiness of the fungi at suitable levels and stops them shrivelling up. Their outer layers contain melanin as a sunblock.

Fungi in even more extreme oceanic environments are harder to study. Those found in deep oceans or below the ocean crust must cope with lack of oxygen and huge barometric pressures. These fungi can't easily be studied because at the sea surface the air pressure is so much lower than they are used to.

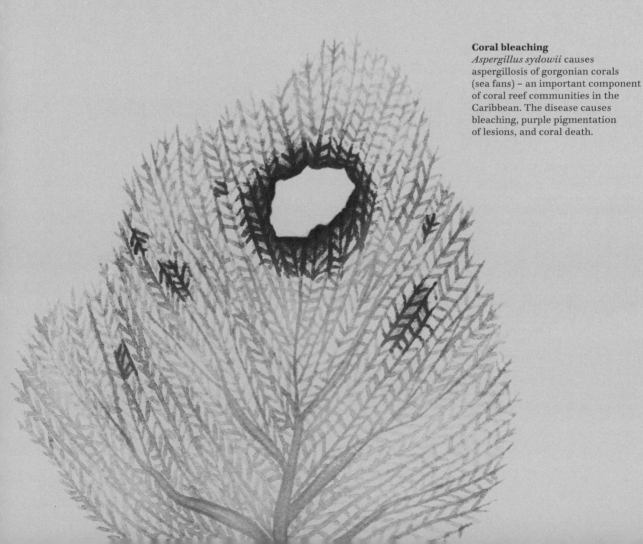

Coral bleaching
Aspergillus sydowii causes aspergillosis of gorgonian corals (sea fans) – an important component of coral reef communities in the Caribbean. The disease causes bleaching, purple pigmentation of lesions, and coral death.

Dunes are composed of sand that has been driven by wind or water into a mound. These are harsh habitats – prone to shifting and reshaping, with low water content and restricted vegetation – but even here, fungi have a home.

SAND DUNES

Dunes are found in both coastal regions and inland, where they form vast dry deserts such as the Sahara and Kalahari in Africa. The size and characteristics of dune systems vary depending on geology, topology, and climate.

SHIFTING SANDS In a coastal dune system (see left), the mobile sand of embryo dunes cannot sustain flowering plants or macrofungi. Although foredunes and "yellow" dunes are also mobile, deep-rooted grasses such as marram and lyme grass provide some stability, supporting some larger fungi, mostly associated with marram grass. Examples include the cup fungus *Peziza ammophila* and the dune stinkhorn (*Phallus hadriani*), which smells and looks similar to its phallic relative found in woodland in temperate Europe.

FIXED DUNES Other macrofungi can be found in the more stable regions further inland, including some common grassland species that are also found away from the coast (pp.140–41), and the dune waxcap (see above right). In these regions in Europe, there is ground-hugging woody vegetation such as willow, where you might find fruit bodies of palefoot saddle (*Helvella leucopus*) with its twisted, wavy dark cap. Fungi that partner with the roots of these dwarf trees are common, and many of them, such as *Cortinarius ammophilus* and *C. fulvosquamosus*, are more or less exclusive to dune slacks, likely thanks at least in part to the high moisture in these regions.

ANATOMY OF COASTAL DUNES

Some coastal dune systems extend for several kilometres and have different "zones". Heading inland from the strand line takes us to embryo dunes, which have mobile sand. Then come foredunes and "yellow" dunes. These are unstable and constantly shift in the wind. Further inland are more stable regions – semi-fixed ("grey") dunes – and stable fixed systems covered with grass or heath. Here, hollows and depressions called "slacks" are susceptible to flooding. Beyond, in temperate regions, is woodland.

A. *Hohenbuehelia culmicola* is a dark brown oyster mushroom that fruits near the base of dead marram grass stems, which it decays. It is found, along with other rare *Hohenbuehelia* species, on semi-fixed sand dunes in Europe. It's cap is 2–7 cm (¾–2¾ in) across.

B. Dune waxcap (*Hygrocybe conicoides*) has a bright cap 2–4 cm (¾–1½ in) across that is silky when dry but greasy when wet. Waxcaps used to be thought to be decayers of dead roots but it now appears likely that they are partners with plant roots or mosses (pp.60–61).

C. Dune brittlestem (*Psathyrella ammophila*) is widespread and common in mobile coastal dunes. The fruit bodies are up to 8 cm (3 in) tall, and appear from late spring to late autumn. The stems extend below the sand surface to mycelium feeding on decaying marram grass.

A.

B.

C.

*With harsh environments that are extremely cold, windy,
and arid (because the water is mostly frozen), it's hard to
imagine anything growing in Antarctica and the Arctic,
yet some fungi call these polar regions home.*

ANTARCTICA AND THE ARCTIC

ANTARCTICA Most of the continent is covered with ice. Although soil temperatures in the maritime Antarctic can reach 20ºC (68ºF) in the summer, the continent holds the record for the lowest temperature on Earth at –89.2ºC (–128.56ºF). Only two flowering plant species can cope with these conditions, and most vegetation consists of mosses. Lichens (pp.70–71) can survive the long periods of drought, and almost 400 lichen species have been recorded – but they grow very slowly. Fungi are also present in soil: between 100 m (330 ft) and 6 km (3¾ miles) of hyphae were detected per gram of soil on Signy Island and the South Orkney Islands. However, only a small number are basidiomycetes. Fruit bodies are therefore rarely seen, but several mushroom species have been found on the South Shetland Islands.

THE ARCTIC At 74º to 81º north latitude in the Arctic Ocean, the Svalbard archipelago has nature reserves over two-thirds of its area, and 60 per cent of the land is covered by glaciers, though these are decreasing due to global warming. Shrubs are encroaching into areas that used to contain mostly mosses and lichens. When scientists warmed small areas, ectomycorrhizal plant species, such as willow, started to grow and mosses decreased. Mushroom-forming fungi are therefore also likely to increase as the climate warms. Most mushrooms in the Arctic are also found in alpine areas but some, such as the milkcap (*Lactarius lanceolatus*), are more common in Arctic regions.

ADAPTING TO THE COLD

The major challenge for fungi in polar regions is stopping ice crystals forming within their cells, as this would cause their cells to rupture. Fungi therefore make antifreeze proteins and chemicals such as glycerol, a bit like antifreeze in car engines. Their cell walls also have a different chemical composition to those of fungi in milder climes, so they do not become brittle in the cold.

Buellia frigida
This lichen grows on rocks at higher
elevations in Antarctica. Conditions
are extremely harsh and it grows
very slowly, extending just half
a millimetre in 100 years.

Arrhenia salina
This fungus is usually only found
on Arctic seashores, but it has also
been reported in the South Shetland
Islands and Antarctica.

Unlike plants, fungi can grow without light; they just need food, water, and an appropriate temperature. Some fungi do require light for their fruit bodies to develop correctly, so those growing underground can often have strange shapes.

CAVES AND MINES

Over 1,000 species of fungi have been detected in underground chambers, but only about 200 of these have macroscopic fruit bodies. Most usually live above ground and are only present in caves by accident – resulting in some weird and wonderful specimens. When it grows in the dark, for example, the train wrecker fungus (*Neolentinus lepideus*) produces stag-horn-shaped fruit bodies rather than its usual mushroom shape. Meanwhile, the fan-shaped turkey tail (*Trametes versicolor*, pp.248), which usually has bands of colours, appears completely white.

TROGLOBITES Fungi that live solely in the dark parts of caves are called troglobite fungi. Among these, basidiomycetes grow on wooden pit props and in animal faeces, such as piles of bat guano. *Antrodia vaillantii* was once so prevalent on wooden pit props that its common name is the mine fungus, though it is less common now metal supports are used.

IMPORTED SPECIES Occasionally, exotic species are imported into mines on timber used for pit props. Fruit bodies of tree mycorrhizal fungi (pp.60–61) are another unexpected sight. Usually, they grow upwards from mycelium near the soil surface, but sometimes the tree roots penetrate into caverns or service conduits, such as sewers. Fruit bodies can then emerge into these underground spaces. Meanwhile, some microfungi are brought in incidentally – on wind, insects, or people's clothing. Those that are insect pathogens outside the cave environment, such as *Beauveria bassiana* and *Isaria*, are often visible because of the masses of spores they produce. These fungi feed on insects in the cave.

Fairy inkcap (*Coprinellus disseminatus*)
Although some fungi look strange below ground, others vary little. This fairy inkcap produces fruit bodies in very large groups on rotting stumps in woodlands. But, like other fungi, it is sometimes found in less usual habitats such as caves and mines, where it can look equally spectacular.

Jack-o-lantern (*Omphalotus olearius*)

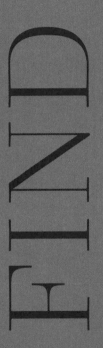

CHAPTER VI

THE WORLD IS FULL OF WEIRD

and wonderful fungal species with names that
are just as intriguing as their looks. This chapter
highlights characteristics from cap and stem
features to microscopic details – and profiles some
fascinating examples of each. Within each section,
the fungi in this chapter have been organized
alphabetically within their form groups.

You don't need to be able to identify fungi to appreciate their beauty or their significance in our ecosystems; however, it is very helpful for determining what is known about a particular fungus and for recording information about it.

IDENTIFYING FUNGI: WHERE TO START

There is no single, "best" way for identifying fungi. The Swedish mycologist Elias Magnus Fries laid the foundations for classifying fungi in the early 19th century by grouping together fungi whose fruit bodies looked similar ("form groups"). However, similar-looking fungi are not necessarily closely related in evolutionary terms, and sometimes fungi that look different are close relatives (p.179). Nonetheless, identifying form groups is still an excellent way to start, and is the approach taken here. With experience, the main forms can be recognized with the naked eye and the help of a hand lens.

The main form groups for basidiomycetes and macroscopic ascomycetes are based on the shape and texture of the fruit body and whether they produce spores on the inside or outside of the fruit body, on gills, tubes, spines, or other surfaces. This chapter, including the individual profiles on pages 180–213, is a starting point for beginners; it is not an identification tool. For detailed identification, consult field guides appropriate to the country or region you're exploring (pp.285–286). Even then, most pocket guides cover just a small proportion of the species found.

Basidiomycetes that produce spores on the outside of fruit bodies:
- stems and caps with gills (pp.180–187), tubes that can be loosened from the rest of the cap (p.190), or veins or smooth undersides of funnel-shaped caps (p.190); spines (pp.188, 191)
- tubes firmly attached to stalked or bracket fungi (pp.192–195)
- smooth, warty, wrinkled, or spiny crusts (pp.196–197)
- rosettes and flattened tongues with smooth surfaces (pp.198–201)
- coral shapes (see right)
- hanging tubes or discs
- rubbery texture (pp.207–208)
- jelly-like feel (pp.206–209)
- rusty or smutty patches on plants (pp.72–73).

Basidiomycetes that produce spores internally:
- puffballs and earthstars (p.204)
- in egg-like structures in containers that look like bird's nests (p.205)
- unpleasant-smelling stinkhorns (pp.202–203)
- basidiomycete truffles: spherical, tuber-like, forming below ground.

The main macroscopic ascomycete form groups (pp.210–213) have:
- ascomycete truffles: spherical, tuber-like, forming below ground (p.227)
- cups whose spore-containing sacs open with a lid
- cups whose spore-containing sacs open without lids
- flasks embedded in a hard tissue (stroma)
- flasks sitting on or in plant tissue
- no fruit body but spore sacs on living leaves and fruits
- tiny spherical fruit bodies on leaves and stems.

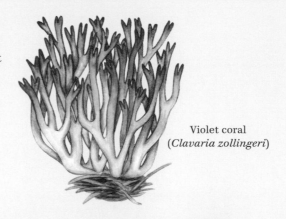

Violet coral
(*Clavaria zollingeri*)

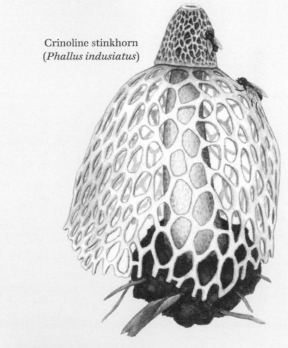

Crinoline stinkhorn
(*Phallus indusiatus*)

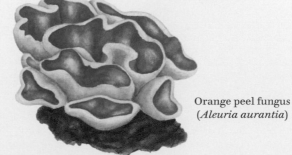

Orange peel fungus
(*Aleuria aurantia*)

*People pick fungi for different reasons, including
for identification and scientific study. Whatever the
intention, it is essential to make sure that it is legal,
ethical, and safe to do so.*

CODE OF CONDUCT FOR PICKING FUNGI

WARNING
Some fungi are extremely
poisonous, even deadly, or can
cause allergic reactions, and
many can make you unwell.
Poisonous and non-poisonous
fungi can easily be confused
(pp.178–179). Take appropriate
care, such as washing your
hands after handling.
Considerable expertise is
needed to identify fungal
species correctly, and this can
take many years to acquire. This
book gives some pointers on
how to start identifying broad
groups of fungi but does not
provide sufficient information
for identifying species. Expert
tuition and guidance can be
gained by attending reputable
courses or joining mycological
societies (pp.285–286).

When it comes to picking fungi and accessing land to do so, the law
varies between countries. Some species are protected by law because
of their rarity and should never be picked or harmed; others are classed
and treated as illegal drugs. In some areas, fungi cannot be picked
without appropriate licences, and there are limitations on how much
can be picked. It is the pickers' responsibility to stay within the law.

Fungi are generally best left growing, but to identify a fungus, you will
often need to collect a specimen (pp.166–167). If you do, keep these
general points in mind.

• Obtain permission from landowners or site managers.
• Like plants, fungi are enjoyed by many people for their natural
  beauty, so leave them for others to see.
• Fungi provide food and habitat for many invertebrates, so pick
  as few as possible.
• Minimize damage to the surrounding habitat, such as vegetation,
  soil, and leaf litter.
• Do not disturb or collect wood.
• Do not pick or damage uncommon, rare, Red List (pp.112–113)
  species, or those that are illegal to collect or possess.

**Scaly-stalked puffball
(*Battarrea phalloides*)**
This rare decomposer fungus has a wide
global distribution in dry sandy soils, but
populations are small and scattered.

Royal bolete (*Butyriboletus regius*)
A rare mycorrhizal species, the
royal bolete is widespread in Europe,
but absent in the north. It is a Red
List species (pp.112–113) in 16
European countries.

COLLECTING FOR SCIENTIFIC RESEARCH

- Collect the minimum required for identification.
- Accurately record site details, so records can be useful to others (pp.166–167).
- Report survey results and interesting findings to landowners or managers.
- Lodge findings in local and national databases, and retain reference or "voucher" specimens.

*Some fungi are easily identified and admired in the field,
but if you want to take your fungi identification skills to
the next level, you'll need to collect specimens for closer
inspection, often using a microscope and sometimes
using simple chemical tests.*

RECORDING, COLLECTING, AND PRESERVING

It's a good idea to note down basic information straight away so you can keep track of interesting findings; a fungus's features may alter even a few hours after you've picked it. Take photographs from different angles (pp.168–169). Then, note down:

- the exact date and location (ideally with grid co-ordinates) of where you found the fungus
- the general habitat, such as conifer woodland, garden lawn, or boggy area
- the species and type of material under or on which it is growing; for example, living plant species, fallen beech twigs, or attached dead oak branches
- any other vegetation close by; remember that tree roots extend for considerable distances so mycorrhizal fungi (pp.60–61), for example, may not necessarily be partners with the tree under which they are fruiting
- whether it appears in a tuft or clump, troop, small group or ring, or near to solitary individuals of the same species.

When collecting specimens, work carefully and record further details.

- Remove specimens carefully with a spatula or small trowel, including the base of the stem and any extensions of the stem below ground.
- Immediately note obvious features that may change or be lost, for example: distinctive odour, hairy cap, colour, or leaking fluid that may follow damage.
- Collect an immature fruit body as well as one or two mature ones, provided there are plenty.
- Cut a wedge from brackets, rather than the whole fruit body.
- Avoid collecting more specimens than you have time to examine.
- Carry specimens in a rigid container with a tight-fitting lid; they will rapidly disintegrate in a plastic bag.
- Keep specimens of different species separate to avoid confusion.

Sheathed wood tuft (*Kuehneromyces mutabilis*)
A decomposer found in clumps on stumps and other wood of broadleaf trees, the sheathed wood tuft has adnate, crowded gills, a ragged ring, and scaly lower stem.

PRESERVING FRUIT BODIES

Specimens can be kept for a few days in a refrigerator. Do not store them with food and label them as not for consumption to avoid any risk of poisoning. They can also be dried gently at 40°C (104°F), then stored in stout paper envelopes or packets in a dry place. These are called vouchers in the scientific world, and are important in case they may be of interest in the future. Larger parts of dried brackets can be stored in cardboard boxes, though they are at risk of burrowing insects, so you may want to use insect or moth repellent. Store spore prints (pp.172–173) with the dried fruit body to help with identification.

If you want to identify and record your fungal finds, it's a good idea to take photographs for reference. Following a few key principles will help you get the most out of your images.

PHOTOGRAPHING FUNGI

1.

A mid-range digital single lens reflex (DSLR) camera with a 200 mm macro lens will enable you to capture plenty of detail for your reference pictures. However, many smartphones have good-quality lenses and can yield excellent pictures with good detail.

2.

A small tripod or camera bean bag will help prevent shaking, so you can use a small aperture and a long exposure to give you a larger depth of field.

3.

Non-direct, natural light under overcast conditions or in hazy sunlight is best. In bright sunlight, the fruit body may need to be shaded, for example using your body. You can use the flash or a sheet of white card as a makeshift reflector to lighten the underside of a cap and reveal gill detail.

4.

To capture the best image of a group of fruit bodies, use the f8 aperture or focus stacking, which will require a tripod and ideally a remote. A shallow depth of field is suitable for single fruit bodies. Image-processing software can help improve sharpness and contrast.

5.

As well as close-ups of the fruit bodies, it is useful to take a context photo showing the habitat of the fungus.

6.

Remove any debris and vegetation obscuring the fruit bodies.

7.

Take several photos from different viewpoints. A photo from above is not sufficient; you also need to capture stem and gill or pore features, for example.

8.

When photographing fruit bodies on the ground, get as low as possible.

IF TAKING PHOTOGRAPHS OF FUNGI WITH A SMARTPHONE, CONSIDER ADDING A CLOSE-UP LENS ATTACHMENT.

9.

If you're removing a wedge of bracket fungus for identification, take a photo of the cut sides to capture the tubes and flesh.

10.

It is useful to photograph buried parts of the fungus, spores deposited on vegetation or lower fruit bodies, exudates (leaking fluid) and changes that occur when damaged, different stages of growth, and other nearby fruit bodies of the same species to show variability.

11.

When possible, photograph fruit bodies with a scale, such as a ruler, for measurement. An object of a known length, such as a specific pen, would also do.

If you want to identify the mushroom you've found, the first step is to look at the cap and stem. Key clues include colour, shape, and texture.

IDENTIFYING MUSHROOMS: CAP AND STEM FEATURES

The appearance and texture of a mushroom's cap and stem can point you in the right direction when it comes to identification. The main signs are listed here. Other features to look out for, which are covered in the following spreads, include spores (pp.172–173), the cells (pp.176–177) and tissues (pp.174–175), and where they are formed.

Marasmius tageticolor
Found in Central and South America, the umbrella-shaped cap, measuring 10–17 mm (½–¾ in) across with distinctive red and white/buff stripes, is supported by a smooth, thin, red-to-dull brown stem measuring 3–4 cm (1¼–1½ in) tall.

CAP: WHAT TO LOOK OUT FOR

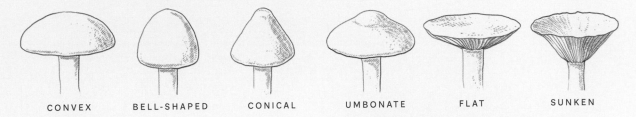

CONVEX   BELL-SHAPED   CONICAL   UMBONATE   FLAT   SUNKEN

Surface colour: Use a standard colour chart (p.286). Colours range from white/grey through pink and red to black; brown is the most common, blue/mauve is uncommon, and green is rare. Colour can vary across the cap, can change with age, and is sometimes darker when wet.

Shape: Young and mature caps vary in shape (see above). The most common are convex, bell-shaped, conical, flat, and sunken/bowl-shaped caps. Note also whether there is a central bump in the cap (called an umbo).

Surface: What does the surface look like? Is it smooth or cracked? Powdery, scaly, shiny, hairy, velvety, woolly? Is it dry, sticky, greasy, or jelly-like?

Peeling cuticle: Observe the cuticle (surface skin). Some can be peeled away to the middle, others only a short distance. This characteristic separates some species, especially *Russula*.

Cap margin: The cap's margin, its outer edge, can be smooth (called entire), or it may be irregularly shaped or split in several places.

Flesh: The internal tissue of the cap is another piece of the identification puzzle. Characteristics to look out for include: thickness from the top of the cap to where the gills emerge; colour, especially changes when it is cut or broken; colour of fluid it exudes when cut or broken (as in *Lactarius* and *Mycena*); and odour.

STEM: WHAT TO LOOK OUT FOR

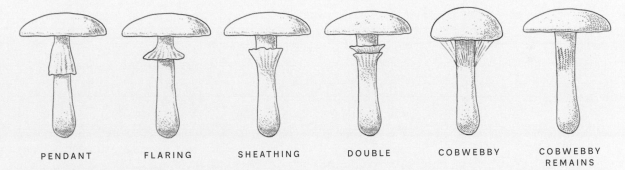

PENDANT   FLARING   SHEATHING   DOUBLE   COBWEBBY   COBWEBBY REMAINS

Attachment to cap: In most mushrooms, the stem is attached centrally beneath the cap. In a few species, the stem is off-centre (excentric) or at the side (lateral).

Shape: Most stems are cylindrical, but a few are slightly flattened or grooved, or may taper slightly upwards or downwards, be club-shaped, or have a bulbous base (pp.50–51). They can be solid, hollow, or partly filled.

Colour and texture: Many characteristics applicable to the cap are important for the stem too, but you should also look out for grooves or wrinkling, and toughness – for example whether it is corky, fibrous, or leathery.

Rings: The remains of the veils that protect the gills when the fruit body emerges (pp.50–51) are often evident on the stem as rings of various shapes and forms: dangling down (pendant), flaring downwards, sheathing upwards, occasionally double, or cobwebby. Sometimes the ring joins the stem loosely (this is called a moveable ring). Parts of the veil may remain as scales on the cap or attached to the outer rim of the cap (pp.50–51).

Volva: Where there was an overall veil (pp.50–51), its remains envelop the base of the stem with parts of the veil protruding as a loose bag (free volva) or closely attached (adherent volva).

As well as being the main way in which fungi spread, spores are important for identification. Observing their colour is simple, but looking at their shape and measuring their size requires a microscope.

IDENTIFYING FUNGI: SPORES AND SPORE PRINTS

SPORE COLOUR All fruit bodies of the same species will have spores of the same colour. Colour is usually the same for different species within the same genus, but it varies in some genera – such as *Flammulaster* and the brittlegills (*Russula*) – so can be important for identification. Describing colour can be subjective, so it is useful to refer to a colour chart (p.286).

Sometimes it can be easy to tell the spore colour – for example, when they have fallen en masse onto vegetation or other fruit bodies. If you want to investigate further, however, it is necessary to make a spore print. This reveals both the colour of the spores and the pattern of the gills, and often looks striking. Always judge the colour in daylight by scraping the spore print into a small heap on clear glass.

SPORE SIZE AND SHAPE Spore size, shape, and surface ornamentation (such as warts) vary between fungal species, making them the key microscopic feature for identification; you'll need a microscope that magnifies by 150 times. Experienced mycologists use special stains to find out whether the spores contain starch (amyloid), which can also help with identification.

MAKING A SPORE PRINT

Cut the cap of a fresh, mature, but not too old fruit body from the top of the stem. Place it gill- or pore-side down on a piece of card. Ideally, the card should be white on one half and black on the other, so the print will show irrespective of its colour. Cover the fruit body with a cup or dish to prevent air currents disturbing the spores. Always make sure you check whether it's appropriate to collect your specimen, and handle it carefully (pp.166–167).

Spore print
Spores drop from the mushroom gills
and fall en masse onto card, making
a print. The pattern they produce
replicates the pattern made by the
gills on the underside of the cap.

Coloured spore prints
Mushroom spores come in a wide range
of colours, from white-cream and pink to
purple-brown and black. Spores are not
necessarily the same colour as the gills
from which they come.

PORES: WHAT TO LOOK OUT FOR

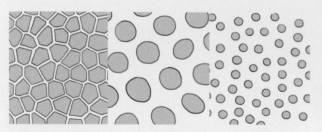

ANGULAR   LARGE CIRCULAR   SMALL CIRCULAR

MERULIOID   CONCENTRIC GILL-LIKE   ELONGATED GILL-LIKE

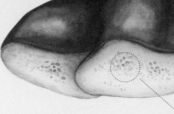

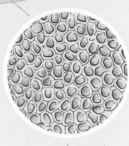

Beef steak (*Fistulina hepatica*)
Spores of this fungus (p.47) are produced on cells in the walls of tubes, which look like small pores on the underside of the fruit body. The pores/tubes of bracket fungi vary in shape and size. Some are even curved and look like gills.

CLOSE-UP OF PORES

GILLS: HOW THEY ARE ATTACHED

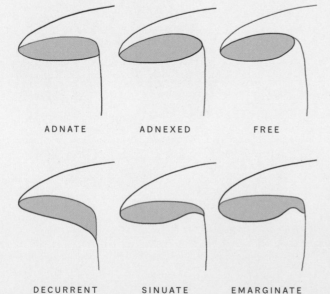

ADNATE   ADNEXED   FREE

DECURRENT   SINUATE   EMARGINATE

The miller (*Clitopilus prunulus*)
Commonly called the miller because of its mealy smell, this fruit body has deeply decurrent gills that are white to start with but turn pinkish as they age.

The gills and pores on the underside of the fruit body are important clues when it comes to identifying fungi. Hyphae in the caps are key features too, but you'll need a microscope to see them.

IDENTIFYING FUNGI: GILLS, PORES, AND HYPHAE

GILLS AND PORES The ridges or "gills" under the caps of some mushrooms can be very revealing. First, check whether they are attached to the stem and how close together they are. Colour is another important identifier: the gills of some fungi change colour as they age, while other species' gills have a different-coloured edge. Gill edges can also be smooth, wavy, or toothed – all of which help identify the fungus. Finally, there are disappearing gills: many species of *Coprinus*, *Coprinopsis*, *Coprinellus*, and *Parasola* have gills that liquify as they mature. Instead of gills, other mushrooms have tubes, revealed as small holes or "pores" under the cap. Look at the length (revealed by cutting a wedge from the cap) and colour of the tubes, number of pores per millimetre length, the shape of the pore – circular or angular – and whether they are visible with the naked eye or using a hand lens. The pores of brackets and flattened fruit bodies have a much wider range of size and shapes, including elongated or maze-like.

HYPHAE Brackets and crusts have three main types of hyphae: skeletal, with thick outside walls and no cross-walls (septa); much-branched binding hyphae with thick walls, and also no septa; and generative hyphae on which the other two types are formed. Fruit bodies with just generative hyphae are called monomitic, those with generative and one other type are dimitic, and those with all three types are trimitic.

When identifying a fungus, fruit body characteristics that can be seen with the naked eye or with a hand lens give good clues. However, microscopic details of the reproductive surface layer are often needed to identify the species correctly.

IDENTIFYING FUNGI: LOOKING AT CELLS

The reproductive or fertile layer (hymenium) – such as the gills, tubes, or surface of crust-like and cup-like fruit bodies – contains the cells that make the spores as well as non-spore-producing cells. A closer look at non-spore-bearing cells can also help. If clamp connections (pp.44–45) are present, it is definitely a basidiomycete. To see these features, you'll need to cut a very thin section of the reproductive layer, and observe it through a microscope.

SPORE-BEARING CELLS In basidiomycetes, the spores (typically four) are made and carried on cells called basidia (see right for examples). An unusual number of spores (two, six, or eight) and, in jelly fungi, shape of the basidia, can help with identification.

In ascomycetes, spores are in sacs. Usually these sacs are cylindrical, but some are more rounded. Important features include: the numbers of spores (commonly eight); whether they are in a straight row; whether the ascus (spore-containing sac) has a single or double wall; and whether spores emerge through a pore with a lid (operculum).

NON-SPORE-BEARING CELLS In fruit bodies, the cells that bear the spores are sometimes interspersed with distinctive non-spore-bearing cells. In basidiomycetes, these cells (called cystidia) are different shapes and sizes depending on the species – from spherical or cylindrical to skittle- or club-shaped (see below for examples). Some are thick-walled, others are thin; some have cross-walls, others don't. Some protrude beyond the spore-bearing cells, resembling tiny spines, and these can even be seen with a hand lens.

In ascomycetes, the cells that do not produce spores (called paraphyses) also have different forms and contents. The tips sometimes protrude above the cells containing spores, forming a protective layer, or they secrete waxy protective substances. Field guides and keys indicate the species for which these characteristics are important.

The shapes of basidia
The spore-bearing cells – basidia – of different groups of basidiomycetes have different characteristic shapes: (a) Lycoperdales (puffballs), (b) Agaricales, (c) Tulasnellales, (d) tuning fork-shaped – Dacrymycetales, (e) Tremellales, (f) transverse septa – Auriculariales.

Cystidia shapes (non-spore-bearing cells)
Cystidia are non-spore-bearing cells found in the fertile layer of fruit bodies of species (not shown here to the same scale). They have a wide range of shapes which are useful characteristics for identification.

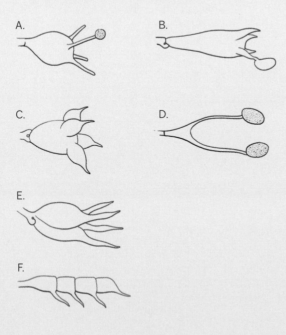

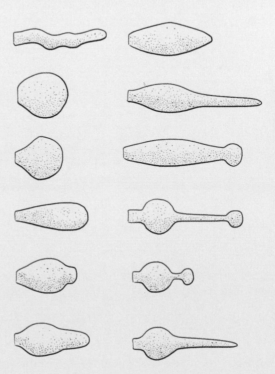

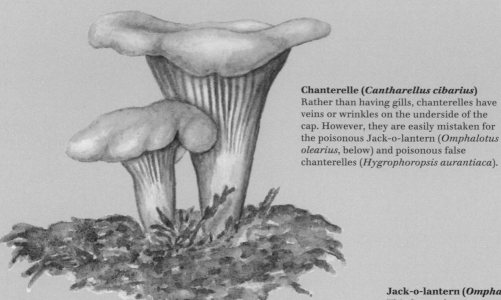

Chanterelle (*Cantharellus cibarius*)
Rather than having gills, chanterelles have
veins or wrinkles on the underside of the
cap. However, they are easily mistaken for
the poisonous Jack-o-lantern (*Omphalotus
olearius*, below) and poisonous false
chanterelles (*Hygrophoropsis aurantiaca*).

Jack-o-lantern (*Omphalotus olearius*)
This fungus has an orange-yellow to
yellow-brown cap with an in-rolled margin
and deeply decurrent, bright yellow gills. This
very poisonous fungus fruits in late summer
and early autumn in central and southern
Europe and North America.

EVOLUTIONARY PROBLEM-SOLVING

Unrelated species have evolved
to have similar characteristics
in a process called convergent
evolution. When species have
to cope with similar issues in
their environment, or need
to perform similar tasks,
they sometimes evolve similar
solutions. This happens in all
kingdoms of life. Fish, marine
mammals, and the now extinct
ichthyosaurs of the Mesozoic
Period, for example, all
evolved a streamlined
body shape.

Some species look very similar to others, at least at first glance. But foragers beware: kingdom fungi has many dangerous doppelgängers in its midst.

DANGEROUS DOPPELGÄNGERS

Some fungi are similar in appearance because they are very closely related and have many characteristics in common. A prime example is the wood-decay fungus chicken of the woods (*Laetiporus sulphureus)*. With its prolific fruit bodies on standing trees (p.131), chicken of the woods was once thought to be a single species, but genetic sequencing has shown that there are at least 11 very closely related but different species that are often difficult to distinguish based on fruit body characteristics alone. On the other hand, other fungi that look similar to each other are not necessarily closely related. Many small brown fruit bodies, for example, are often just identified as "little brown mushrooms", unless someone is particularly interested in studying their biology.

Looks can be deceptive. Fungus doppelgängers can be extremely dangerous as some poisonous species look very similar to edible ones (see left). Some fungi are deadly poisonous (pp.96–97). Never eat a fungus that is growing in the wild unless you are 100 per cent certain that its identification as an edible fungus is correct. Even experts with many years' experience can be deceived.

AMANITA VIROSA
Destroying angel

FRUIT BODY This deadly poisonous fungus has a distinctive volva (pp.50–51) at the base and a ring high on the stem. Remnants of the veil are rarely seen on the cap. The gills are white, free, and crowded.

SIZE Cap: 5–10 cm (2–4 in) across. Stem: 0.5–2 cm (¼–¾ in) wide, 9–15 cm (3½–6 in) long.

SPORES Spherical and contain starch, producing a white spore print.

HABITAT AND ECOLOGY The fungus is mycorrhizal with conifers and broadleaf trees. Fruit bodies usually appear in summer and early autumn.

DISTRIBUTION Widespread and common in the northern hemisphere.

AGARICS

Agarics are basidiomycetes. Fungi in this form group usually have a stem and vertical gills on the underside of the cap. They can also be subdivided into additional form groups. The way in which the gills attach to the stem (p.174) is a key feature for identification. In many agarics, the gills are protected during fruit body development by one or two veils, leaving remnants on the stem of mature fruit bodies (pp.50–51), although some have no protection. Spore colour is another important identification characteristic. Different species have spores of different colours, ranging from white, cream, and yellow to brown, black, and – more rarely – lilac or green. Agarics can be found across the globe. With more than 3,000 species in temperate Europe alone, it is not surprising that this form group counts decomposers, mycorrhizas, and plant pathogens among its members.

ARRHENIA CHLOROCYANEA
Verdigris navel

CLITOCYBE NEBULARIS
Clouded funnel

FRUIT BODY The verdigris navel has a small, turquoise to blue, funnel-shaped cap with very distant, white-to-grey gills that are deeply decurrent (p.174).

SIZE Cap: 0.5–2 cm (¼–¾ in) across. Stem: 0.75–2.5 cm (⅓–1 in) long.

SPORES Cylindrical to ellipsoidal in shape, producing a white spore print.

HABITAT AND ECOLOGY This decomposer fungus is found in disturbed sandy or gravelly soil with mosses, lichens, and occasionally liverworts. Fruit bodies are seen all year round, particularly March to April, appearing alone or in small groups.

DISTRIBUTION Common in Europe and found in parts of North America.

FRUIT BODY This large, fleshy mushroom has a convex cap that becomes funnel-shaped with a slightly in-rolled edge. The gills are slightly decurrent (p.174). Mature fruit bodies can be parasitized by the piggyback rosegill (*Volvariella surrecta*).

SIZE Cap: 5–25 cm (2–9 in) across. Stem: 2–3 cm (¾– 1¼ in) wide, 6–12 cm (2⅓–4¾ in) tall.

SPORES Ellipsoidal and smooth, creating a creamy white spore print.

HABITAT AND ECOLOGY A leaf litter decomposer, the fungus usually appears in large rings or arcs on leaf litter in broadleaf and conifer woodland (p.136). Fruit bodies appear from late summer to early winter.

DISTRIBUTION Widespread in Europe and North America.

COPRINOPSIS PICACEA
Magpie inkcap

FRUIT BODY Young fruit bodies have an unpleasant smell. The white gills are crowded and free or adnate (p.174), but deliquesce into a black liquid with age (p.175).

SIZE Cap: 3–7 cm (1¼–2¾ in) across. Stem: 7–12 cm (2¾–4¾ in) long, 0.5 to 1.5 cm (¼–⅗ in) across.

SPORES Smooth, black, and ellipsoidal.

HABITAT AND ECOLOGY These decomposer fungi often appear in broadleaf woodland but also in shaded grassland with woody debris on alkaline soils. Fruit bodies appear from late spring to late autumn.

DISTRIBUTION Widespread and common in Europe and North America.

GLIOPHORUS PSITTACINUS
Parrot waxcap

FRUIT BODY The gills are thick, waxy, and widely spaced, narrowly attached to the stem (adnexed, p.174) and are initially greener closer to the stem, fading to pale yellow as the fruit body matures.

SIZE Cap: 2–4 cm (¾–1½ in) across. Stem: 4–8 mm across, 4–6 cm (1½–2⅓ in) tall.

SPORES Ellipsoidal and smooth, producing a white spore print.

HABITAT AND ECOLOGY Parrot waxcaps are found in small groups on unfertilized grasslands (pp.140–141), lawns and cemeteries, roadside grass verges, and woodland clearings. Fruit bodies appear in summer until early winter.

DISTRIBUTION Widespread in Europe, north and central America, and Japan.

HEBELOMA RADICOSUM
Rooting poison pie

LACTARIUS DELICIOSUS
Saffron milkcap

FRUIT BODY This fruit body has a distinctive almond smell. Its gills (p.174) are crowded, adnexed to free, cream to red-brown, covered by a partial veil when immature. The stem, which tapers towards the base and extends below ground to its food source, has a membranous woolly ring.

SIZE Cap: 5–10 cm (2–4 in). Stem: 7–12 cm (2¾–4¾ in) long.

SPORES Ellipsoid to almond-shaped, somewhat warty, producing a dark brown spore print.

HABITAT AND ECOLOGY Often found near conifer stumps, this fungus is associated with the below-ground latrines of small mammals. The fruit bodies appear from early summer to late autumn.

DISTRIBUTION Widespread but uncommon in Europe and North America.

FRUIT BODY Initially convex with an in-rolled margin, the caps flatten with age and become depressed in the centre, appearing almost vase-like, with irregular green markings. When cut, the gills ooze a bright orange latex that occasionally becomes redder. The stems are hollow.

SIZE Cap: 6–20 cm (2⅖–8 in) across. Stem: 1.5–2 cm (⅖–¾ in) across, 5–8 cm (2–3 in) long.

SPORES Ellipsoidal with ridges, producing pale pink to buff prints.

HABITAT AND ECOLOGY Saffron milkcaps are ectomycorrhizal (pp.60–61) with pine. The fruit bodies appear from late summer to autumn.

DISTRIBUTION Widespread in Europe, with introductions to other countries.

MACROLEPIOTA PROCERA
Parasol mushroom

MYCENA HAEMATOPUS
Burgundydrop bonnet or bleeding mycena

FRUIT BODY Initially spherical, the large cap flattens as it matures and has concentric brown scales and a dark brown central umbo. It has a large, double-edged ring high on the stem. The gills (p.174) are free, crowded, and white or cream, sometimes with a pink tinge.

SIZE Cap: 10–25 cm (4–9 in) across. Stem: up to 30 cm (12 in) tall, 1–1.5 cm (⅖–⅗ in) across.

SPORES Ellipsoidal and smooth with thick walls, producing creamy white prints.

HABITAT AND ECOLOGY A decomposer in grasslands, the fruit bodies appear summer to late autumn.

DISTRIBUTION Widespread and common in temperate Europe and North America.

FRUIT BODY The caps are weakly bioluminescent (pp.124–125), and the gills are adnate (p.174) and whitish. Both the cap and stem, which is hollow, bleed a brownish-red "latex" when damaged. Many *Mycena* species are slender and fragile, but this species is much larger.

SIZE Cap: 1–4 cm (⅖–1½ in) long. Stem: up to 10 cm (4 in) long and 2–3 mm across.

SPORES Ellipsoidal and contain starch (p.172), creating a white print.

HABITAT AND ECOLOGY A decomposer fungus on large dead wood of broadleaf trees, its fruit bodies grow in small groups or clusters and appear in late summer to winter.

DISTRIBUTION Widespread and common in Europe, North America, and Japan.

PANAEOLUS PAPILIONACEUS
Petticoat mottlegill

PAXILLUS INVOLUTUS
Brown roll rim

FRUIT BODY The dome- or bell-shaped cap never flattens. It has pale, tooth-like projections around the margin. The gills are adnate (p.174) and pale grey-brown with white edges that blacken with age.

SIZE Cap: 2–4 cm (¾–1½ in) across. Stem: 6–12 cm (2⅖–4¾ in) long.

SPORES Lemon-shaped, smooth, and opaque, producing a black print.

HABITAT AND ECOLOGY Petticoat mottlegills occur on dung and well-manured soil in grassland. Fruit bodies appear from spring to early winter.

DISTRIBUTION Widely distributed and very common in Europe and North America. Occasionally recorded on other continents.

FRUIT BODY The fruit body is deadly poisonous. Its pale gills are easily loosened from the cap flesh. The gills and stem turn rusty brown with age or if damaged.

SIZE Cap: 5–12 cm (2–4¾ in) across. Stem: 6–12 cm (2 ⅕–4¾ in) long, 0.8–1.2 cm (⅓–½ in) across.

SPORES Ellipsoidal and smooth, producing a yellow-brown to olive-grey print.

HABITAT AND ECOLOGY This fungus is ectomycorrhizal (pp.60–61) on broadleaf trees and conifers, especially on poor, acidic soils. Fruit bodies appear from summer to autumn.

DISTRIBUTION Widespread and common in Europe, North America, and occasionally recorded on other continents.

PHOLIOTA SQUARROSA
Shaggy scalycap

RHODOTUS PALMATUS
Wrinkled peach

FRUIT BODY These large fruit bodies, with dry caps, grow in clusters and are an important food for red squirrels. They have crowded, adnate (p.174) gills, which start yellowish and turn rust-brown.

SIZE Cap: 2–16 cm (¾–6 ⅓ in) across. Stem: 4–12 cm (1½–4¾ in) long, up to 1.5 cm (⅔ in) wide.

SPORES Bean-shaped spores, producing a brown spore print. It is sometimes confused with *Armillaria*, but this has white spores.

HABITAT AND ECOLOGY This wood decomposer fruits at the base and higher up on trunks and stumps of broadleaf trees and sometimes on conifers, causing white rot. Fruit bodies appear from late summer to winter.

DISTRIBUTION Common in Europe and North America.

FRUIT BODY The cap is rosy pink with an in-rolled margin, becoming peach-coloured and flatter as it matures. It has adnate gills (p.174).

SIZE Cap: 5–10 cm (2–4 in) across. Stem: 3–7 cm (1¼–2¾ in) long, 1–1.5 cm (⅖–⅗ in) across.

SPORES Spherical and covered in fine warts, producing a pale pink print.

HABITAT AND ECOLOGY Wrinkled peach is a decomposer of dead trunks and branches, particularly elm and beech. Fruit bodies appear in late summer and autumn.

DISTRIBUTION Found in Europe and eastern North America, it is now rare and local due to the decline in elm populations resulting from Dutch elm disease (pp.78–79).

HERICIUM ERINACEUS
Lion's mane, bearded tooth, or pom-pom mushroom

FRUIT BODY These are white to cream-coloured and roughly spherical. Their pointed spines emerge from the same place and dangle downwards.

SIZE Up to 30 cm (12 in) across.

SPORES Roughly ellipsoidal or slightly more rounded and containing starch, producing a white spore print.

HABITAT AND ECOLOGY It is a wood-decay fungus, causing white rot in the trunks of standing living, dead, and fallen beech and less commonly oak trees. Annual fruit bodies are produced from late summer to early winter.

DISTRIBUTION Although found in North America, Europe, and parts of Asia, it is widespread but rare in most of Europe.

FLESHY BASIDIOMYCETES WITH PORES, SPINES, OR VEINS

BOLETOIDS Species in this form group are all closely related, though some close relatives take different forms. They have tubes on the underside of the caps instead of gills. The tubes usually come away from the cap flesh easily. The colour of the tubes, the surface of the cap, and how the colour changes when damaged are important characteristics for identification. The spores are usually brown and are made on the inside walls of the tubes. A few are parasites on other fungi (pp.102–103), but most form ectomycorrhizal partnerships with the roots of trees (pp.60–61), and are often selective about the tree species.

FUNNEL-SHAPED WITH A SMOOTH OR VEINED UNDERSIDE The stalked, fleshy fruit bodies of chanterelles and similar fungi form a funnel shape, with much of the upper surface inside the funnel, while the lower surface of the cap – where spores are formed – is on the outside. Unlike other funnel-shaped fruit bodies, spores form on veins rather than on gills, but it is easy to confuse the two, which could have deadly consequences (pp.178–179). Most form mycorrhizal partnerships with tree roots, but a few are decomposers.

HYDNOIDS Hydnoid mushrooms have spiny or tooth-like undersides, on which the spores are produced. They are often not closely related but have relatives in other fruit body form groups. Some are wood-decay fungi, while others form mycorrhizal partnerships with tree roots. Some crust fungi also have teeth (pp.196–197).

LECCINUM SCABRUM
Brown birch bolete

CRATERELLUS CORNUCOPIOIDES
Horn of plenty

FRUIT BODY The clay- to buff-coloured cap is first felty then smooth as it matures. The light-coloured stem has greyish-brown or black scales.

SIZE Cap: 5–15 cm (2–6 in) across. Stem: 7–20 cm (2¾–8 in) long, 2–3 cm (¾–1¼ in) wide. Tubes: 1–2 cm (⅕–¾ in) long.

SPORES Ellipsoidal to spindle-shaped, producing an olive-brown spore print.

HABITAT AND ECOLOGY It forms mycorrhizal partnerships (pp.60–61) with birch. Fruit bodies appear from late spring to late autumn.

DISTRIBUTION Widespread and common in the northern hemisphere.

FRUIT BODY This funnel-shaped fungus has a grey, slightly powdery, and almost smooth outer layer with fine veins (equivalent to the gills on other mushrooms) running down the stem.

SIZE Cap: 4–8 cm (1½–3 in) across.

SPORES: Ellipsoidal and smooth, creating a white spore print.

HABITAT AND ECOLOGY It forms mycorrhizal partnerships (pp.60–61) with roots of beech and sometimes other trees. Fruit bodies appear in summer to late autumn, and into the new year in warmer places.

DISTRIBUTION Found in Europe, North America, parts of Asia and south-eastern Australia, it is geographically widespread but found in localized spots, where it is often abundant.

PHELLODON NIGER
Black tooth

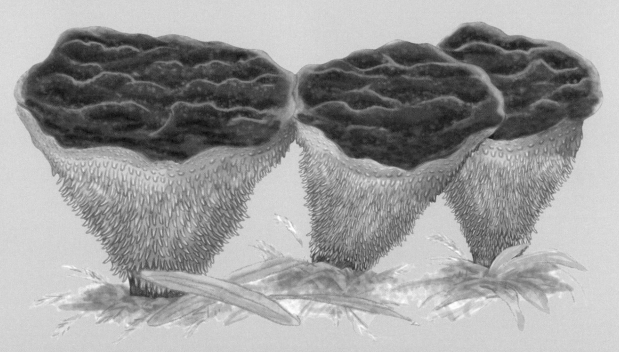

FRUIT BODY This hydnoid produces rough, flattish fruit bodies, often with shallow dips. The lower surface has blue-grey spines up to 3 mm long.

SIZE Cap: 3–8 cm (1¼–3 in) across. Stem: 2–5 cm (¾–2 in) long, 1–2 cm (⅓–¾ in) wide.

SPORES Ellipsoidal to more rounded, with tiny (about 0.5 μm) spines, producing a white print.

HABITAT AND ECOLOGY It forms mycorrhizal partnerships (pp.60–61) with roots of conifers. Fruit bodies appear from summer to late autumn in northern Europe but for a longer period further south.

DISTRIBUTION Widespread but rare in Europe and eastern North America, Central America, Japan, and south-eastern Australia.

CERIOPORUS SQUAMOSUS
Dryad's saddle or pheasant's back mushroom

FRUIT BODY Individual caps with short, wide stems often grow in overlapping clusters and smell like watermelon. The lower surface is white to cream with tube-like pores that are irregular ovals, 1–2 mm wide.

SIZE Cap: 8–30 cm (3–12 in) across. Stem: up to 5 cm (2 in) long, up to 4 cm (1½ in) wide.

SPORES Smooth and ellipsoidal, creating a white print.

HABITAT AND ECOLOGY Dryad's saddle causes white rot in the heartwood (pp.130–131) of living and dead broadleaf trees, as well as on stumps and large logs. Fruit bodies appear from spring to late autumn.

DISTRIBUTION Widespread and common in Europe, Asia, and North America.

POLYPORES

Over 1,000 species of polypore have been described by science so far. They all belong to the phylum Basidiomycota, though not all are closely related. Some are perennial, but most are annual. Annual polypores can be separated into four slightly overlapping groups based on whether they: have a stem; form in clusters; are completely flat (resupinate); or are soft or tough but not hard brackets, some of which produce tiers of brackets. In all polypores, spores are formed inside tubes that point downwards. The mouths of the tubes (called pores) are circular, or angular, or resemble a labyrinth, or are long and narrow, resembling gills, depending on the species (pp.174–175). Almost all fungi that form polypore fruit bodies are wood decomposers, producing white or brown rot. Several are being studied for possible medicinal properties.

DAEDALEA QUERCINA
Mazegill

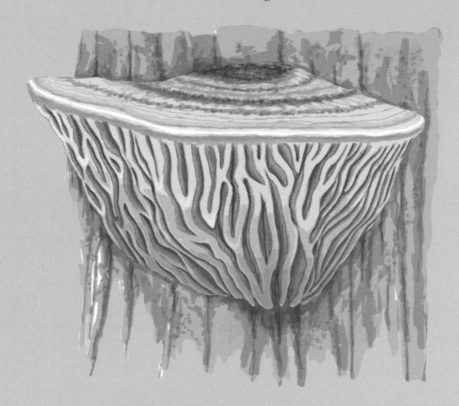

FRUIT BODY The common name of this tough bracket is inspired by its pores, which look like a maze of thick gills that are 1–3 cm (⅖–1¼ in) deep. It is trimitic, meaning the fruit body is made up of three different types of hyphae (p.175).

SIZE Cap: 6–20 cm (2⅖–8 in) across, 2–5 cm (¾–2 in) thick.

SPORES Shed from late summer to autumn, the spores create a white print.

HABITAT AND ECOLOGY Mazegills cause brown rot (pp.120–121). They grow mainly on felled or fallen logs and stumps of oak or sweet chestnut. Fruit bodies can be seen all year.

DISTRIBUTION Widespread in Europe, west Asia, and North America.

PHAEOLUS SCHWEINITZII
Dyer's polypore, dyer's mazegill, or velvet top

FRUIT BODY These shelves grow around vegetation, such as blades of grass, which can be seen poking through the fruit body. The tube layer is a greenish-yellow when young and reddish-brown when older.

SIZE Cap: 15–25 cm (6–9 in) across, 2–5 cm (¾–2 in) thick. Stem: less than 6 cm (2⅓ in) long and 3–5 cm (1¼–2 in) across.

SPORES Ellipsodial, smooth, and don't contain starch, producing a white to pale yellow print.

HABITAT AND ECOLOGY This fungus causes brown rot of wood (pp.120–121). It is usually found on the roots or at the base of trunks of living or dead conifer trees. Fruit bodies are more common in summer and autumn.

DISTRIBUTION Widespread and common. Native in Europe, Asia, North America, but introduced into South Africa, Australia, and New Zealand.

CHONDROSTEREUM PURPUREUM
Silverleaf fungus

FRUIT BODY Silverleaf fungus crusts are purplish in colour, smooth, with a pale, hairy, greyish upper side if a small bracket-like cap is present (not shown here). While fresh, specimens are hard to mistake for anything else, but old fruit bodies fade and look similar to other species. At a microscopic level, spindle-shaped cystidia (pp.176–177) project from the spore layer.

SIZE Patches often about 3–4 cm (1¼–1½ in) wide, but can merge to form larger patches.

SPORES Cylindrical with rounded ends, the spores do not contain starch.

HABITAT AND ECOLOGY Silverleaf fungus grows on the wood of many broadleaf trees, causing the often-fatal silver leaf disease, especially on cherry and plum trees. Fruit bodies can be found all year.

DISTRIBUTION Widespread and common in the northern hemisphere, eastern Australia, and New Zealand.

CRUSTS

The fungi that form crust fruit bodies are basidiomycetes. There are about 500 named species in temperate parts of Europe. Many are unrelated to each other. The fruit bodies are flat (resupinate), often forming on the underside of wood, and/or have a narrow, slightly protruding cap. Some have distinctive macroscopic characteristics, such as colour or shape, which makes them easy to identify with the naked eye, but most need to be studied with a microscope for correct identification. Some have spines, teeth, warts, others veins or folds, or smooth cottony, or smooth waxy surfaces. Most crusts are white-rot decomposers of wood, but some are brown rotters (pp.120–121). Finally, some crusts – particularly those in the Thelephorales order – form mycorrhizas with plant roots (pp.60–61).

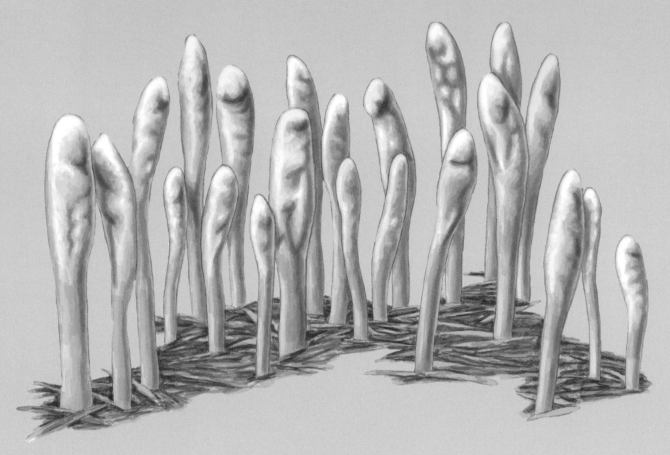

CLAVARIADELPHUS LIGULA
Strap coral

FRUIT BODY Initially pale yellow-cream, the narrow club- or spoon-shaped fruit bodies darken with age. They are initially smooth, apart from hairs at the base, becoming wrinkled. The fertile layer covers the surface.

SIZE Each club: 3–12 cm (1–4¾ in) tall, up to 2 cm (¾ in) across at the thickest point.

SPORES Smooth, elongated-elliptical, forming a white to pale yellowish spore print; they do not contain starch.

HABITAT AND ECOLOGY A decomposer fungus found in coniferous forests with mycelium at the base binding together needles and woody debris. Fruit bodies are formed in late summer and autumn.

DISTRIBUTION Widespread in Europe, North America and Asia.

ROSETTES AND CORALLOID BASIDIOMYCETES

ROSETTES AND FLATTENED TONGUES With fewer than 30 species in temperate Europe, this is a small grouping. They are not closely related, but they are all shaped like fans, single rosettes, or multiple rosettes with numerous folds. Spores are produced on the smooth under or outer surface. Species range in size from fans that are just 1–2 cm (⅓–¾ in) tall to multiple rosettes that are 40 cm (16 in) wide. *Thelephora* species form mycorrhizas with plant roots (pp.60–61), but most others are decomposers or parasites of mosses.

CORALLOID These fruit bodies are all club-shaped. One of the simplest forms in kingdom fungi, club-shaped fruit bodies have evolved many times in lots of different species. There are about 200 species in temperate Europe, many of which are unrelated. The simplest are single, small (under 5 mm/¼ in tall and a fraction of a millimetre wide), thin, unbranched, or slightly branched structures. Some species grow upright, while others grow in a more random or even downward orientation. Others form large clubs that are 7 cm (2¾ in) or taller and several centimetres wide. Yet others are highly branched clusters of thin clubs, resembling coral. The relationship between many of these fungi and plants is still not understood: some are decomposers, others form ectomycorrhizas (pp.60–61), and yet others are plant parasites.

THELEPHORA PALMATA
Stinking earth fan or fetid false coral

FRUIT BODY This branched, coral-shaped fungus
has slightly flattened tufts that branch repeatedly
from a central stem. Whitish initially, it turns grey
to brown with age. It smells like rotten cabbage.

SIZE Reaches up to 7 cm (2¾ in) tall.

SPORES Brown, angular, and warty.

HABITAT AND ECOLOGY Stinking earth fan
is a mycorrhizal partner of conifer trees (pp.60–61).
Its fruit bodies appear from late summer to the
end of autumn.

DISTRIBUTION Widespread and fairly common in
Europe, North America, Japan, and Australia.

TYPHULA ERYTHROPUS

FRUIT BODY One of the larger *Typhula* species, this simple, unbranched club has a red-brown hairy stalk, which can be seen with a good hand lens, and a white cylindrical head on which spores are produced. It often arises from a red-brown sclerotium (p.289).

SIZE About 2–3 cm (¾–1¼ in) tall.

SPORES Contain starch.

HABITAT AND ECOLOGY *Typhula erythropus* is a decomposer of the leaves and leaf stalks of broadleaf trees. Fruit bodies appear in autumn.

DISTRIBUTION Widespread and common in Europe, but less common further north.

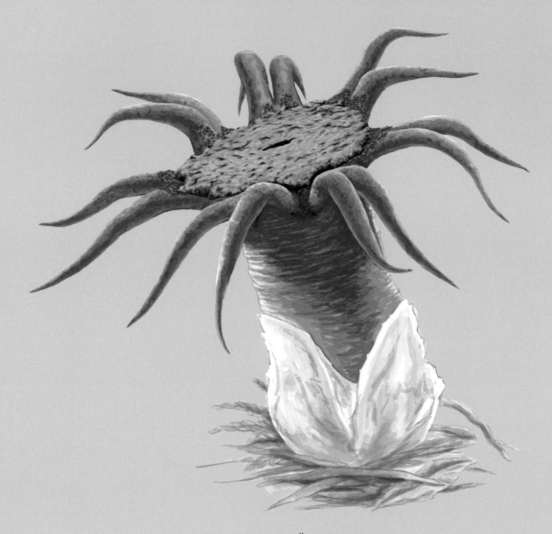

ASEROË RUBRA
Starfish fungus or anemone stinkhorn

FRUIT BODY The hollow stem emerges from an "egg". Six to twelve "arms" or "tentacles" radiate from the top of the stem, and a smelly, dark olive-brown to black mass of spores forms on the stem end of the tentacles.

SIZE Fruit body: about 10 cm (4 in) tall, arms up to 4 cm (1½ in) long. "Egg" (p.203): 3 cm (1¼ in) across.

SPORES Ellipsoidal, dark-coloured, and smooth.

HABITAT AND ECOLOGY A decomposer of plant litter, wood chips, and mulch, starfish fungus is common in gardens, in Eucalyptus, and semi-evergreen to evergreen forests.

DISTRIBUTION Common and widespread in eastern Australia and New Zealand but has spread widely.

BASIDIOMYCETES WITH ENCLOSED SPORES

STINKHORNS The fruit bodies form within a spherical or egg-shaped structure, which is partly or completely underground. When it opens, a dark, slimy mass of spores is raised above the ground. The fruit bodies smell of rotting flesh that attracts insects (pp.56–57).

PUFFBALLS AND EARTHSTARS Most puffballs and earthstars are Agaricales, which is the same taxonomic order as mushrooms with gills. The enclosed ball- or onion-shaped fruit bodies contain a powdery mass of spores that puff out due to pressure from a raindrop or falling twig (pp.56–57). Some genera have no stem, others have short, wide stems, and a few sit on a tall stem. In earthstars (mostly *Geastrum* species), the outer layer cracks and each part bends backwards, raising the inner ball up onto "stilts" (p.204). In others, such as *Scleroderma*, the thick outside layer splits when it matures. Most genera, including *Lycoperdon*, are decomposers, but some, such as *Scleroderma*, are mycorrhizal root partners (p.60).

BIRD'S NEST FUNGI All bird's nest fungi are Agaricales, and they decompose wood and plant litter. Their smooth, transparent spores are produced on the usual spore-producing cells (basidia) but within packages (called peridioles) inside cup-shaped structures that look like eggs in a bird's nest – hence the name of the form group. Spore packages are usually splashed out by drops of rain.

ASTRAEUS HYGROMETRICUS
Barometer earthstar or false earthstar

FRUIT BODY This earthstar has a felty spore sac. When humidity is high, the sac is raised on 6–15 speckled "stilts" (rays), which fold back over in drier conditions.

SIZE Spore sac: 1–3 cm (⅛–1¼ in) across, and up to 8 cm (3 in) from tip to tip when expanded.

SPORES Approximately spherical with tiny warts, producing a reddish-brown spore print.

HABITAT AND ECOLOGY This earthstar is a mycorrhizal root partner (pp.60–61) with oak and pine trees. Fruit bodies can be seen all year, often after rain in late summer and autumn.

DISTRIBUTION Widespread in Europe, North America, Asia, and Australasia but not in cold regions.

NIDULARIA DEFORMIS
Pea-shaped bird's nest

SPHAEROBOLUS STELLATUS
Artillery or cannonball fungus

FRUIT BODY This bird's nest fungus has an outer cushion-shaped sac. Each sac contains flattened, brown "eggs" that are white inside and contain the spores.

SIZE Fruit body: 0.5–1 cm (¼–⅓ in) across. "Egg": 0.5–2 mm across, 0.3 mm thick.

SPORES Smooth, transparent or white, ellipsoidal, and do not contain starch.

HABITAT AND ECOLOGY Pea-shaped bird's nest is a decomposer on wet, rotting wood of conifer and broadleaf trees. Fruit bodies appear summer to early winter.

DISTRIBUTION Widespread but not common in Europe, North America, and New Zealand.

FRUIT BODY Growing in groups, this tiny bird's nest fungus resembles pale-coloured spheres or egg shapes. These crack open to reveal a pale brown- to orange-coloured "ball" that contains the spores. When mature, the ball is shot away.

SIZE 1–3 mm diameter.

SPORES Thick-walled and clear.

HABITAT AND ECOLOGY Artillery fungus is a decomposer on well-decayed wood, dead herbaceous material, and old dung. Fruit bodies are seen from early summer to early winter.

DISTRIBUTION Widespread and common in Europe, North America, Australia, and Japan.

EXIDIA GLANDULOSA AND *E. NIGRICANS*
Witches' butter and Warlock's butter

FRUIT BODY Both fruit bodies are similar and very often confused. *Exidia nigricans* looks brain-like when there are masses together, while *E. glandulosa* often also grows in tight groups and has a warty lower side. The fruit bodies can dry out to crisp-like crusts, which take up water when it rains and start to produce spores again.

SIZE Individual fruit bodies: 1–2 cm (⅓–¾ in) across. Masses: 10 cm (4 in) or more.

SPORES Sausage-shaped, not containing starch, and producing a white spore print.

HABITAT AND ECOLOGY Both species are decay fungi on recently dead attached or fallen wood, on a range of broadleaf trees – *E. glandulosa* especially on oak, *E. nigricans* more often on beech, ash, and hazel. Fruit bodies are seen all year, though they are often at their best during autumn and mild winters.

DISTRIBUTION Very common and widespread in Europe and North America.

OTHER BASIDIOMYCETE FRUIT BODIES

RUBBERY FUNGI These species are all closely related, as members of the order Dacrymycetales. Fruit bodies are usually yellow in colour and gelatinous or rubbery in texture. Spore-producing cells are shaped like tuning forks (pp.176–177). Asexual spores bud off the sexual spores. All are wood decayers causing brown rot (pp.120–121).

HANGING DISCS OR TUBES (CYPHELLOID) Most of the fungi that produce tiny tube-, cup-, or bell-shaped fruit bodies that point downwards are closely related and belong to the order Agaricales – many of which form mushroom-shaped fruit bodies. Unlike mushrooms, however, cyphelloids have lost their gills during evolution. The smallest species are about 0.2 mm across and the largest may be 10 mm across.

JELLY FUNGI The jelly-like fungi fruit bodies vary in colour and shape. Some species have basidia with long projections that bear the spores (sterigmata, p.289) and septa (p.16), which are visible through a microscope. Most are either decomposers or parasites of plants or fungi.

CALOCERA VISCOSA
Yellow stagshorn

FRUIT BODY These rubbery fungi are bright orange-yellow, with greasy, antler-like branches that are branched two to four times.

SIZE Up to 10 cm (4 in) tall.

SPORES Slightly curved to sausage-shaped, containing oil droplets but no starch and producing a white spore print.

HABITAT AND ECOLOGY This wood-decay fungus grows on conifer wood that is often buried beneath the soil surface. Fruit bodies appear throughout the year but especially in autumn.

DISTRIBUTION Widespread in Europe and North America.

TREMELLA MESENTERICA
Yellow brain fungus or star jelly

FRUIT BODY This jelly fungus is gelatinous when wet. It turns much paler if there has been prolonged rainfall.

SIZE Individual fruit bodies are 2–8 cm (¾–3 in) across.

SPORES Basidiospores do not contain starch, and produce a white spore print.

HABITAT AND ECOLOGY Fruit bodies are found on dead wood, or breaking through the bark of broadleaf trees, but it does not decay wood – it is a parasite of other fungi (pp.102–103). Fruit bodies can be seen all year, but not often in late spring or early summer.

DISTRIBUTION Very common and widespread on all continents except Antarctica.

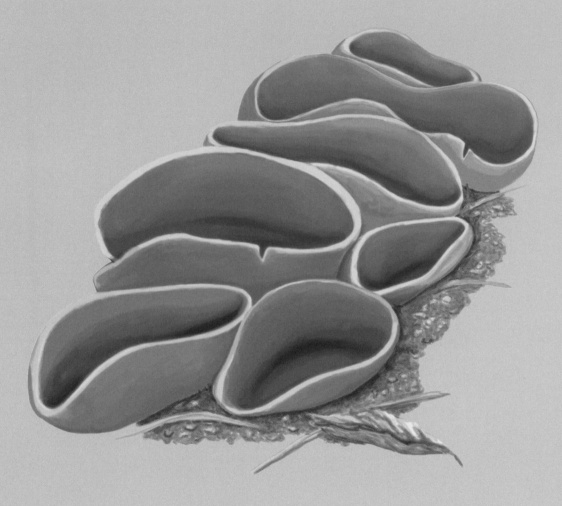

ALEURIA AURANTIA
Orange peel fungus

FRUIT BODY The large, cup-shaped fruit body expands and becomes more irregular on maturity, developing wavy margins (edges) and often splitting. Appearing in groups, fruit bodies are attached to the soil by mycelial cords.

SIZE Each cup is 3–10 cm (1¼–4 in) across.

SPORES Ellipsoidal with ornamentation, containing two oil drops, and producing a white spore print.

HABITAT AND ECOLOGY Orange peel fungi often grow on gravelly soil, on and beside disturbed paths in parks and forests. Fruit bodies appear from summer through to midwinter.

DISTRIBUTION Widespread and common in Europe and North America.

ASCOMYCETES: CUP, SADDLE, AND FLASK FUNGI

Fungi in this group are all in the phylum Ascomycota (pp.48–49). Some are ectomycorrhizal, others are parasitic or decomposers of dead plant material and animal remains. Ascomycetes produce spores in sacs called asci (p.49), often surrounded by sterile hyphae (paraphyses, p.177). The fruit bodies of this group vary from microscopic to macroscopic, and they range in shape from simple cups, flasks, or ball-like structures to more elaborate ears, goblets, saddles, and honeycombs.

Ascomycetes are often called "spore shooters" as many propel their spores from mature fruit bodies. For some, the spore-producing surface (hymenium) of the fruit body is fully exposed, facilitating spore release. Depending on the genus or species, the hymenium ranges in colour from white, grey, black, or brown to violet, pink, yellow-orange, or red. It can therefore offer clues for identification, as can spore size, colour, and ornamentation (whether spores are warty, spiny, and so on). Mycologists also often use microscopic features to identify ascomycetes, such as the presence or absence of a small lid (operculum) at the top of each ascus, the presence of a double ascus wall, and the arrangement of paraphyses within the fruit body.

HELVELLA MACROPUS
Felt saddle

MICROSTOMA PROTRACTUM
Rosy goblet

FRUIT BODY The outer surface is hairy and dull, while the upper, spore-producing surface is greyish and shiny. It is closely related to the false morel and therefore likely to be poisonous.

SIZE Cup: 1–4 cm (⅓–1½ in) across. Stem: 1–5 cm (⅓–2 in) long.

SPORES Ellipsoidal to spindle-shaped, smooth but sometimes with fine warts on the surface, producing a white spore print.

HABITAT AND ECOLOGY These fruit bodies can be found singly or in groups on rich soil or rotten wood, often in broadleaf forests. Fruit bodies appear from early summer to early autumn.

DISTRIBUTION Widespread and common in Europe, North America, and Asia.

FRUIT BODY These goblet-shaped fruit bodies are produced in clusters and joined to a common, deeply rooted stem. The cup is ball-shaped with a central opening, eventually becoming disc-shaped with a ragged margin.

SIZE Cups: 4 cm (1½ in) tall and 2.5 cm (1 in) across.

SPORES Large with thick walls, producing a white spore print.

HABITAT AND ECOLOGY Rosy goblets appear in clusters in woodland in spring. They grow on buried wood on rich soils that are often periodically wet.

DISTRIBUTION Though rare, they are found in Europe, North America, and northern Asia.

OTIDEA ONOTICA
Donkey ears or hare's ear

XYLARIA POLYMORPHA
Dead man's fingers

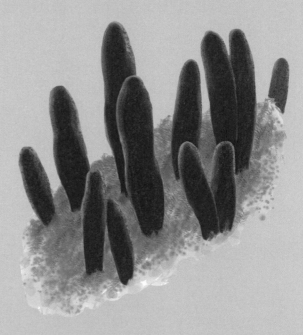

FRUIT BODY The apricot-orange outer surface is dull and slightly downy (floccose), while the inner hymenium is smooth and sometimes spotted or with a slight pink tinge.

SIZE Cups: 3–9 cm (1¼–3½ in) tall, 1.5–4 cm (⅗–1½ in) across. Stem: 1 cm (⅓ in) tall.

SPORES Its spores are white, smooth, and ellipsoidal, with two oil drops.

HABITAT AND ECOLOGY Donkey ears are found on the floor of coniferous and broadleaf forests. Variously reported as a decomposer and ectomycorrhizal fungus, its fruit bodies appear from spring to autumn.

DISTRIBUTION Widespread and common in temperate Europe and North America.

FRUIT BODY The tiny, flask-shaped fruit bodies are embedded in finger-like structures (stromata). When sliced open, they reveal a white, fleshy interior with black, flask-shaped fruit bodies around the surface. Each flask contains numerous asci (p.49).

SIZE Stromata: 3–8 cm (1¼–3 in) in length, 1–3 cm (⅖–1¼ in) diameter.

SPORES Smooth, dark brown, and spindle- or banana-shaped, producing a black print.

HABITAT AND ECOLOGY This wood-decay fungus emerges from the ground and appears at the base of deciduous tree stumps or from buried wood. Fruit bodies appear mostly in summer and autumn.

DISTRIBUTION Widespread and common in Europe, North America, and elsewhere.

Pink oyster mushroom (Pleurotus djamor)

GROW

HUMANS HAVE KNOWN about the nutritional value and health benefits of some fungi for millennia, but it is only relatively recently that mushrooms have been cultivated on a commercial scale. Discover the historic culinary uses of fungi – including the gourmet truffle – and learn how they grow.

Mushrooms have been celebrated in spiritual practices for tens of thousands of years, but the earthy, umami tastes of some of the edible varieties have also appealed to humans for centuries.

HISTORIC CULINARY USE OF MUSHROOMS

12,000YA

By studying human dental material found in caves at El Mirón in the mountainous region of Cantabria, Spain, scientists were able to show that certain fungi – such as some species of boletes – were consumed by Palaeolithic people in that region between 18,000 and 12,000 years ago. This is one of the earliest records of mushroom consumption by humans other than for spiritual practices.

4,500YA

Ancient Egyptian hieroglyphs show that mushrooms and truffles were consumed as far back as 4,500 years ago. According to legend, mushrooms were "the sons of the gods" and were thought to be associated with immortality. As a consequence, only royalty were permitted to consume them.

THE EGYPTIANS REALIZED THAT THE MICROSCOPIC FUNGUS YEAST COULD BE USED TO MAKE BREAD

ENOKITAKE GROWN
IN NATURE

ENOKITAKE GROWN
IN THE DARK

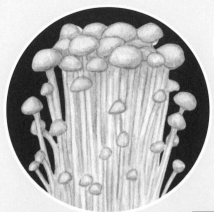

2,200YA

The jelly ear fungus (*Auricularia cornea*) was first cultivated in China over 2,200 years ago. Also known as "cloud ear" or "hairy wood ear", its crunchy texture is still valued in Chinese cuisine. Meanwhile the slightly sweet-tasting enokitake (*Flammulina filiformis*) has been grown in Japan and China for centuries. The fungus is now commonly cultivated in the dark (see image above).

2,000YA

Greek horticulturalists began cultivating mushrooms on animal dung over 2,000 years ago. They were also able to culture certain fungi on a small scale, such as the nutty-flavoured velvet pioppini (*Cyclocybe aegerita*), a mushroom that was also grown by the Romans and is still available in European markets today.

400YA

A serendipitous discovery brought mushroom cultivation to Paris in the 1600s where white button mushrooms (*Agaricus bisporus*) were first grown in the palace gardens at Versailles. In the mid-1800s, the catacombs and old quarries of Paris became the cultivation ground for these agarics. However, this ended in the 20th century when the Metro was constructed.

Thanks to their fleshy texture and rich umami flavour profiles, certain mushrooms have been eaten for thousands of years. Thirty or so varieties are now cultivated for use in cuisines around the world, and for good reason.

WHY EAT CULTIVATED MUSHROOMS?

Many popular mushrooms are grown on an industrial scale (pp.220–223), including the cultivated mushroom (*Agaricus bisporus*) and oyster mushrooms (*Pleurotus* species). Their texture and versatility make them valuable in vegetarian and vegan diets, while mycelium is used as a meat alternative (pp.272–273). Yeast, a microscopic single-celled fungus, is used to produce spreads and nutritional yeast flakes.

NUTRITIONAL VALUE Mushrooms' nutritional value is often underestimated. Low in fat and sodium, they contain reasonably high levels of fibre and vitamins and minerals, such as B vitamins, as well as selenium, copper, and potassium (levels vary from species to species). Selenium has antioxidant properties, while copper is important in body processes, such as producing red blood cells, and potassium is crucial in cell functioning. Including a pack of cultivated mushrooms in a dish can provide a similar amount of potassium to a banana.

Mushrooms and other fungi contain a substance in their cell membranes called ergosterol, which is similar to cholesterol found in animals. However, since ergosterol is not used by our bodies in the same way as cholesterol, mushrooms are cholesterol-free foods.

VEGAN VITAMIN D

Vitamin D is essential for many processes in the body. We get vitamin D predominantly via exposure to sunlight, but also through some foods. Fungi produce vitamin D through ergosterol (see right). Exposing mushrooms to ultraviolet light increases vitamin D levels by converting ergosterol into a form of vitamin D. Other food sources of vitamin D include oily fish and egg yolks, but mushrooms are the only natural source of the vitamin for those who follow a vegan diet.

SOME CONTAIN VITAMINS AND MINERALS
THOUGHT TO BE IMPORTANT TO HEALTH

SOME HAVE HIGH
VITAMIN D CONTENT

SOME CONTAIN THE
ANTIOXIDANT SELENIUM

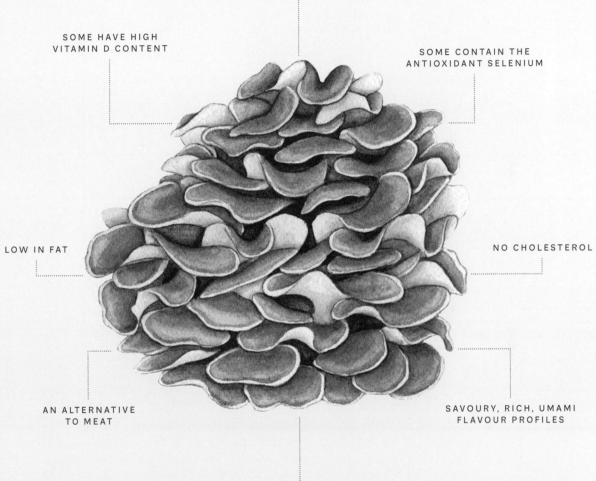

LOW IN FAT

NO CHOLESTEROL

AN ALTERNATIVE
TO MEAT

SAVOURY, RICH, UMAMI
FLAVOUR PROFILES

SOME CONTAIN COMPOUNDS THAT HAVE
ANTI-INFLAMMATORY PROPERTIES

*Cultivated mushrooms are relatively easy to grow
compared to most food crops, and the market for both commercially
and home-grown mushrooms is booming. When growing any mushroom,
three factors are key: substrate, environment, and spawn.*

GROWING
MUSHROOMS

CAUTIONARY NOTES

Growing mushrooms can be rewarding and fun, but to avoid any risk of growing non-edible mushrooms, be sure to obtain fungal spawn or complete growth kits from a reputable supplier. When growing mushrooms outdoors, there is also a risk that spores from a non-edible species from the natural environment could replace an edible species. If there is the slightest doubt, seek advice from an expert. Some people are sensitive to mushrooms, so the first time you try a new type, only sample a small amount. As with plant pollen, fungal spores can be allergens, so avoid inhalation and ensure that mushrooms are grown in well-ventilated places. It is sometimes possible to buy strains that produce fewer spores.

Mushrooms that are decomposers (p.39) can be grown locally on waste materials and often at scale, which reduces the distance they need to travel to reach the consumer ("food miles") and therefore their impact on the environment. Other cultivated fungi can break down wood; the white-rot fungus shiitake (*Lentinula edodes*), for example, is often cultured on logs. Mycorrhizal fungi (pp.60–61), such as truffles, are more challenging to grow both on a small scale and commercially because they require a living partner tree.

THE KEY TO GROWING EDIBLE MUSHROOMS When cultivating mushrooms, two of the most important factors to consider are establishing the correct environmental conditions for growth and fruit body formation, and choosing the correct food source (in this case, the substrate) for the fungal species. The main difference between species is that button mushrooms require pre-digested substrate, whereas the other edible mushrooms do not. The third most important factor is to obtain good-quality mushroom spawn (a sort of starter culture). This consists of a pure culture of fungal spores or mycelium that is mixed with and allowed to colonize sterilized grain or wooden dowels. The spawn is then added to the relevant substrate and the fungus continues to grow, producing mycelium and eventually mushrooms.

Pink oyster mushroom (*Pleurotus djamor*)
This is a popular mushroom grown from commercially available kits. It produces abundant pink, fan-shaped fruit bodies that are stunningly beautiful.

Most people associate "edible mushrooms" with the white button mushroom Agaricus bisporus. *In fact, white button mushrooms, chestnut or cremini mushrooms, and the larger portobello mushrooms are all* Agaricus bisporus, *so they are all grown in the same way.*

GROWING WHITE BUTTON MUSHROOMS

Agaricus bisporus has been cultivated in much the same way since the technique was developed in France in the 1800s, and the mushrooms are now grown commercially in many countries around the world. You can even grow the button mushroom at home in a matter of weeks using a kit, which consists of mycelium-colonized compost, requiring just the addition of the casing layer (see right), which is included in the kit, and water.

THE SUBSTRATE All mushrooms need a substrate (or compost), a material in which the mycelium can grow. Unlike some of the other commercially grown mushrooms, *Agaricus bisporus* needs a predigested mushroom compost. To achieve this, the raw material substrate (usually straw-bedded horse manure) is allowed to ferment in the presence of naturally occurring microorganisms. The resulting substrate is rich in nutrients for fungal growth and mushroom production.

BUTTON MUSHROOM SPAWN Mushroom spawn is prepared by adding *Agaricus bisporus* mycelium to sterilized millet grain; this is manufactured at scale by specialist companies. Mushroom growers obtain their spawn directly from these suppliers.

HOW BUTTON MUSHROOMS ARE GROWN COMMERCIALLY

1.
For *Agaricus bisporus*, wheat straw-bedded horse manure is the preferred substrate, with other ingredients such as seed hulls, nitrogen (often added as chicken manure), and gypsum (to stabilize the acidity and prevent the straw from sticking together).

2.
Heat-loving microorganisms naturally present in the substrate begin to ferment it, altering the nutritional content of the compost, making it more favourable for mushroom growth and less favourable for other fungi and bacteria.

3.
The bulky yellow compost is watered and, as it reaches the required temperature, it is turned regularly over a 2–3 week period, until it becomes a dark, dense mixture smelling of ammonia.

4.
It is then spread on to pallets and pasteurized to remove unwanted fungi, insects, nematodes, and other pests, while minimizing the loss of beneficial microorganisms, which help with the removal of ammonia.

5.
The compost temperature is reduced. The mushroom spawn is inoculated into the compost and temperature, humidity, and carbon dioxide levels are controlled (p.53) to maximize mycelium growth.

6.
Once the mycelium has grown through the compost (after 7–10 days), it is covered with a "casing layer" of peat and limestone to stimulate immature mushrooms (primordia or pins) to form (p.45).

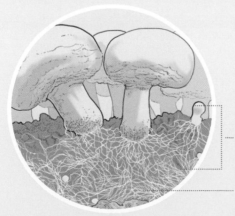

7.
Two or three weeks later, mushrooms develop and can be picked, stimulating more growth for later harvests. The whole commercial process takes about 14 weeks.

CASING LAYER SHOWING MYCELIAL CORDS AND MUSHROOM PRIMORDIA (PINS)

MUSHROOM COMPOST COLONIZED BY *AGARICUS BISPORUS* MYCELIUM

*Some edible mushrooms, such as oysters and shiitake,
will grow happily on wood-based substrates that require no pre-digestion.
There are many kits available for growing these mushrooms at home,
but care should be taken, especially if growing outdoors.*

GROWING OTHER EDIBLE MUSHROOMS

OYSTER MUSHROOMS These fungi are some of the fastest to produce mushrooms, and are a good starting point for beginners. Spawn can be purchased from specialist suppliers, and the mushrooms can grow on a variety of substrates. Kits come in sterilized bags, often with prepared substrate inoculated with spawn. They may require soaking in water and a cold shock to stimulate fruit body formation. Make slits in the substrate bags to allow exposure to air, and mist the bags regularly with water to maintain humidity levels. Depending on the chosen method, oyster mushrooms can be harvested from between 2 and 6 weeks and can be grown in a shed, basement, or even in the airing cupboard. Care should be taken to avoid inhalation of spores (p.220).

Commercially, oyster mushrooms are grown in temperature-controlled rooms. The spawn is added to sterilized substrate of wheat straw and wood-based materials, and incubated for 2–3 weeks to allow mycelium to develop. This is then soaked in water and requires a cold shock and air exposure to stimulate fruiting within 7–10 days. Temperature and humidity are carefully monitored to optimize mushroom production.

SHIITAKE MUSHROOMS Grown for centuries in Asia, these wood-decay fungi prefer to fruit on sawdust-based substrates or logs (preferably oak, birch, and alder). Mushroom spawn usually comes on grain or as colonized wooden dowels that can be inserted directly into freshly cut and drilled hardwood logs. On logs, shiitake usually fruit twice a year, in spring and autumn. In sawdust bags, about 2–3 weeks after the colonized grain has been added, make slits and soak in water for 1–2 days. The first mushrooms appear in 1–2 weeks. Commercially, both log impregnation and sawdust-based substrates are used, and temperature and humidity are optimized for mushroom production.

Shiitake mushroom
Native to East Asia, the Shiitake mushroom has been cultivated for centuries and widely used in traditional Chinese medicine. Today, Shiitake mushrooms are grown commercially across the globe and are a common sight in many supermarkets.

Truffles are often referred to as the Mozart of the mushroom world – and for very good reason. This culinary delight is the most expensive edible mushroom, but growing it is not for the faint-hearted.

GROWING TRUFFLES

Truffles are subterranean fungi, producing their fruit bodies underground, in close association with trees. The fungus is ectomycorrhizal (pp.60–61) and partners with various trees, including oak, beech, larch, and hazel, depending on the fungal species. In nature, mature truffles give off an aroma that can be detected by squirrels, rabbits, meerkats, and other animals, which dig them up, enabling these underground fruit bodies to spread their spores. Traditionally, humans have enlisted the help of truffle pigs and dogs to sniff them out.

SLOW GROWERS As truffles need a plant partner, they will not grow in compost, sawdust, or logs like many other edible mushrooms. Many online suppliers offer truffle-impregnated trees for purchase; however, it can take a decade or longer to establish an ectomycorrhizal partnership of sufficient maturity to allow this ascomycete fungus to produce its truffle fruit bodies – and a further decade before a reliable harvest is achieved. Some growers impregnate soil with truffle mycelium before planting tree seedlings. Others harvest seedlings from the brûlé, the area around the truffles' partner tree. Saplings with mycorrhizal connections thrive in this zone, so harvesting and replanting them should result in a tree with natural fungal connections.

COMMERCIAL TRUFFLE FARMING There is a small but growing number of truffle farms in the USA, Australia, and Europe. The soil needs to be soft and friable, well-drained, and mildly alkaline. A temperate climate is also important. Growing at scale involves either inoculating sapling roots with truffle spores or growing seedlings in truffle-impregnated bagged soils. At 2–5 years old, the trees are planted into suitable ground. It is then likely to be 7–10 years before any truffle is harvested, and up to 20 years before productivity reaches a commercial level.

TRUFFLE FARMING ON
A COMMERCIAL SCALE

1.

Tree saplings are potted into soil that contains truffle spores or mycelium.

2.

The saplings are grown in pots for 2–5 years. They are repotted into larger pots containing additional soil and truffle inoculum until the saplings develop into young trees.

3.

Trees are planted out in rows in soil that is well-drained, friable, and slightly alkaline – conditions that suit the establishment of truffle associations.

4.

Trees are pruned annually and the soil is irrigated. It will be 7–10 years before the first truffles begin to appear, and full productivity can take up to 20 years.

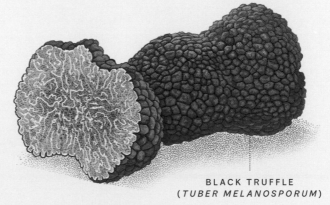

BLACK TRUFFLE
(*TUBER MELANOSPORUM*)

5.

Truffles are harvested from just below the soil surface, taking care not to damage tree roots. Truffle dogs may be used to sniff them out.

6.

Although cultivating truffles is time intensive, the rewards are high, with the most prized truffle varieties costing over £3,000 per kilo.

Fly agaric (*Amanita muscaria*)

CELEBRATE

CHAPTER VIII

MUSHROOMS HAVE FEATURED

in art, folklore, music, and rituals for as long as humans have walked this planet. Discover how mushrooms inspired superstitions and folklore, stories of fungi that were thought to have higher powers, the myths and legends behind fungal names, and even their ability to produce music.

Often erupting from the ground as if by magic, mushrooms have captivated humans for centuries. It is hardly surprising, then, that they have a deep-rooted history in folklore.

MUSHROOM MYTHS AND LEGENDS

FAIRY RINGS Fairy rings (pp.136–139) are steeped in mycological folklore. The overnight appearance of these peculiar sights has fuelled many myths. In Austria and parts of France, fairy rings were thought to be places of magic and sorcery, made by dragons who breathed fire or swished their tails to create the scorched circle. English folk tales tell of faeries who came to dance in circles in celebration of the rain, causing the grass to fade within the ring's centre. According to old English superstition, entering the ring will bring about 100 years of stupor, while bathing on the dew inside a fairy ring will cause a young maiden's skin to erupt in blemishes and spots. So well-known were these myths that fairy rings are mentioned by Prospero in Shakespeare's *The Tempest*.

FOOD OF GODS In ancient Greece, mushrooms were thought to symbolize fertility. Their importance in Greek mythology is hinted at in a stone carving of the Greek goddess Persephone being reunited with her mother Demeter after the daughter's period of captivity in the underworld. Mushrooms feature in the joyous scene, possibly symbolizing rebirth or as a nod to the hallucinogenic properties of certain mushrooms, which some believe to have influenced ancient Greek religious rites.

Suggestive stinkhorns
It is said that Beatrix Potter did not have the courage to paint it, and Charles Darwin's daughter collected and destroyed it for fear of it humiliating the servants. It was even drawn upside down in Victorian texts. Yet the fruit body of the stinkhorn (*Phallus impudicus*), presumably because of its appearance, is mistakenly thought by some to be an aphrodisiac.

Royal baking
English legend has it that, when he was taking refuge in Somerset's marshlands during a surprise invasion by the Danes, King Alfred fell asleep while baking cakes. The hard, shiny, black mature fruit bodies of the fungus *Daldinia concentrica* resemble burnt cakes, hence one of its English common names: King Alfred's cakes.

Auspicious fungi
Also known as *reishi* (meaning "divine" or
"spiritual mushroom" in Japanese) or *lingzhi*
(meaning "tree of life" in Chinese), this bracket
fungus (*Ganoderma lucidum*) was thought
to bring good health and luck to those who
were associated with it.

REVERED FUNGI The dark, glossy, fan-shaped bracket fungus (*Ganoderma lucidum*) remains the most widely depicted mushroom in China, Japan, and Korea. Its image appears on temple wall carvings, statues, paintings, and drawings, and it is even carved in wooden sceptres (or *ruyi*). In Beijing's Forbidden City and Summer Palace, the emperor's official sceptre, as well as furniture and other structures, all carry carvings showing the fruit body of this fungus.

THE SANTA TOADSTOOL The fly agaric (*Amanita muscaria*, pp.239, 241–243), with its striking red cap and white spots, is perhaps the most iconic mushroom in mycological folklore – and its festive colouring may not be a coincidence. According to legend, the origins of Santa Claus may stem from Siberian shamanic practices involving this particular fungus.

Shamans would dry out fly agaric mushrooms by hanging them from pine trees, and then hand them out as gifts in winter. The story goes that Santa's suit originated from the red deer pelts and red-and-white spotted garments worn by the shamans, and that the flying reindeer came from hallucinations experienced after eating the mushrooms.

Fly agaric appears under birch and pine as if by magic – just like gifts under the Christmas tree – and red-and-white mushroom decorations are a common sight in Europe during the festive season.

FATAL FUNGI

It's commonly thought that Napoleon died of a stomach ulcer while in exile on the island of St Helena, but it has also been suggested that his death involved a fungus. Some believe that the French leader inhaled toxic chemicals produced when the bathroom mould *Scopulariopsis brevicaulis* caused the Scheele's green paint pigment on his walls to break down, releasing toxic arsine gas.

From prehistoric cave drawings to 20th-century stamps,
fungi have been widely represented in art – as symbols of spirituality,
as valued foods, and for their diverse shapes, colours, and forms.

MUSHROOMS IN ART

ANCIENT MUSHROOM ART The prehistoric cave paintings at Tassili in North Africa, Neolithic stone-carved mushroom gods in Guatemala, and mushroom inscriptions on Indian Magadha Janapada coins showing mushroom shapes alongside the tree of life, all testify to our ancestors' desire to share their reverence of mushrooms through art.

RENAISSANCE TO BAROQUE Paintings of fungi, usually brackets on trees or mushroom clumps on the forest floor, became more widespread during the Renaissance, highlighting an increased awareness of fungi. Fungi were also a source of inspiration during the still-life era of the Baroque period, notably in Italy. Painters such as Bartolomeo Arbotori often included the Caesar's mushroom (*Amanita caesarea*, p.97) and the penny bun (*Boletus edulis*). Art from Asia at this time often depicted the lingzhi or reishi mushroom (*Ganoderma lucidum*, pp.232–233).

THE ROMANTICS AND BEYOND The Romantic period saw an explosion of interest in the natural world. Amateur mycologists began to collate records through drawings, paintings, dried specimens, and field notes. Some of the most exquisite drawings are by Anna Maria Hussey, and by the author Beatrix Potter who painted over 300 fruit bodies and lichens, mostly from Scotland and the Lake District.

MUSHROOM STAMPS

The first stamps to show mushrooms were produced in Romania 100 years after the world's first stamp, the Penny Black, was issued in 1840. The first fungus to appear on a British stamp was *Penicillium* in 1967 to commemorate the success of the antibiotic penicillin (pp.252–253). Today, mushroom stamps are collected by philatelists worldwide.

Renaissance fungi
Giuseppe Arcimboldo's 1573 painting *Autumn*
shows autumnal fruits, vegetables, and mushrooms
in the form of a human head, where the ear is
clearly a gilled mushroom.

*From music-making spores to myco-wood violins that
replicate the quality and pitch of the finest string instruments
in the world, fungi are more musically tuned than you
might think.*

MUSHROOM MUSIC

MUSICAL SPORES Mushrooms release spores into the atmosphere, allowing fungi to reach new habitats (pp.54–55). Each mushroom's pattern of spore release is unique, and this pattern can be used to create music. This natural process is normally hidden from view. However, if a laser beam is directed beneath the cap of a mushroom, it can visually capture the release of microscopic spores as they pass through the beam of light. By designing custom-built lasers linked to music-generating software, sound designer Yann Seznec and mycologist Patrick Hickey in Edinburgh, Scotland, created a system that generated sounds to represent the spores as they were released from the mushroom and fell through the laser beam. The result was a delicate, tinkling kind of mushroom music.

FUNGAL VIOLINS When wood is selectively decayed by the white rot fungus *Physisporinus vitreus* and *Schizophyllum commune*, the resulting material can be crafted into a "myco-wood" violin, an instrument with enhanced acoustic properties on a par with the world-renowned Stradivarius family of string instruments. Stradivarius violins were made by the artisan craftsman Antonio Stradivari during the Baroque period of the 17th to the mid-18th century, using European spruce from high in the Italian Alps. Trees growing in this period had been exposed to a "little ice age" and produced wood that was more chemically uniform (homogeneous) than trees grown in warmer climates, giving them a better acoustic sound when crafted into musical instruments. It is this homogeneity, combined with the exceptional craftmanship of Stradivari, that make Stradivarius instruments so unique. And fungi can be used to create a similar effect. "Myco-wood" violins therefore have the potential to provide talented musicians with a high-quality instrument at a more affordable price.

Schizophyllum commune
Controlled degradation of wood by the
white-rot fungi *Physisporinus vitreus*
and *Schizophyllum commune* (above)
can produce wood with similar acoustic
properties to wood used to make violins
during the Baroque period.

*In many fairytales and children's books, **fairies** are depicted sitting on top of the red-and-white **fly agaric** mushroom, or living inside* **mushroom houses**.

From Anglo-Saxon poetry to 19th-century children's stories, authors and poets throughout history have taken inspiration from fungi and some of their key attributes.

FUNGI IN LITERATURE

GLOWING FUNGI The oldest surviving poem in English, *Beowulf* contains a reference to fungal bioluminescence (pp.124–125.): "o'er it the frost-bound forest hanging, sturdily rooted, shadows the wave. By night is a wonder weird to see, fire on the waters." During the Middle Ages, glowing mushrooms or glowing wood – here, probably reflecting off water – were considered otherworldly.

FUNGAL DECAY Death and decay caused by fungi was an inspiration to 19th-century English writers such as Percy Bysshe Shelley and Charles Dickens. In Dickens's *Great Expectations*, for example, Miss Havisham's decaying home and wedding feast seem representative of the decay at the heart of the upper classes' decadence, in contrast to the extreme poverty of the era. A much more light-hearted reference to decay appears in the 20th century with *Fungus the Bogeyman* by Raymond Briggs. In this picture book, much of the humour centres on the "dirtiness" of the main characters: Fungus, his wife Mildew, and his son Mould.

HALLUCINOGENIC FUNGI The hallucinogenic properties of certain fungi have also been depicted. Lewis Carroll's *Alice's Adventures in Wonderland*, written in 1865, is a notorious example. Alice meets a large blue caterpillar, who is sitting on top of a mushroom and smoking a long hookah. When he leaves the mushroom and disappears into the long grass, he says, "One side will make you grow taller, and the other side will make you grow shorter." This scene evokes the psychedelic properties of the fly agaric (*Amanita muscaria*) and the liberty cap (*Psilocybe semilanceata*; pp.260–261) and their mind-altering influence on the body.

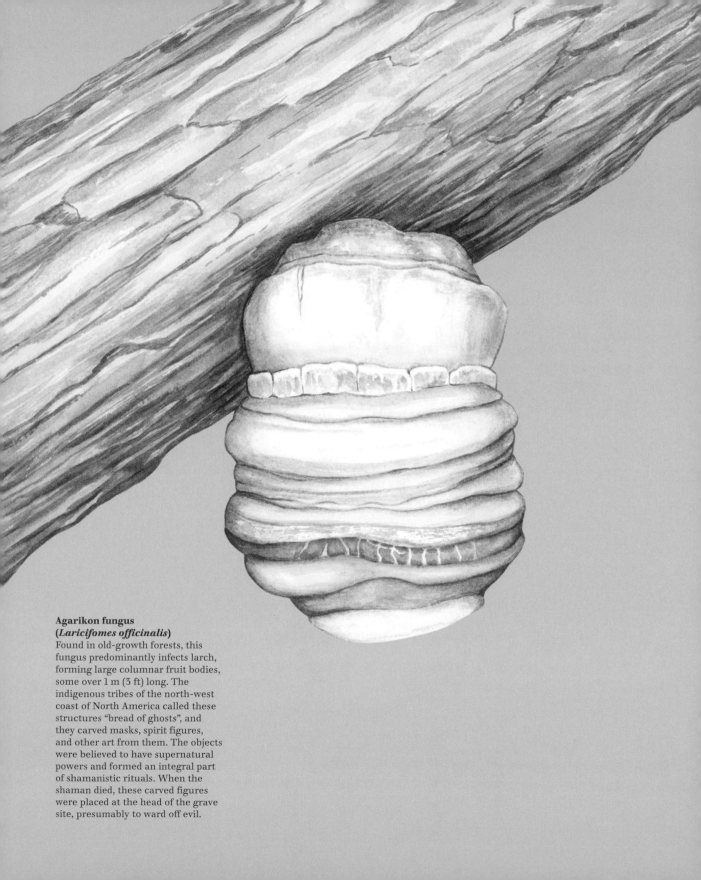

Agarikon fungus
(***Laricifomes officinalis***)
Found in old-growth forests, this fungus predominantly infects larch, forming large columnar fruit bodies, some over 1 m (3 ft) long. The indigenous tribes of the north-west coast of North America called these structures "bread of ghosts", and they carved masks, spirit figures, and other art from them. The objects were believed to have supernatural powers and formed an integral part of shamanistic rituals. When the shaman died, these carved figures were placed at the head of the grave site, presumably to ward off evil.

Ancient mushroom symbolism dating back several centuries can be found in many regions of the world. Some mushrooms were even worshipped, forming the basis of age-old spiritual ceremonies and rituals.

MUSHROOMS IN CEREMONIES AND RITUALS

CAVE ART Some of the earliest records of humans using mushrooms in spiritual practice date back over 9,000 years. Prehistoric art in the Tassili caves in Algeria shows dancing figures holding mushrooms that emit light rays that strike the head of each dancer, perhaps representing their hallucinogenic effects. Drawings also seem to show gods with mushrooms emerging from their bodies.

STONE MUSHROOMS Enigmatic mushroom-shaped stones thought to be over 2,500 years old have been found throughout northern Central America. The stones are carved into the shape of two hallucinogenic mushrooms: the liberty cap (*Psilocybe semilanceata*, pp.260–261) and the fly agaric (*Amanita muscaria*, pp.242–243). The stalks depict a god, the cap a hat, and smaller mushrooms are sometimes carved around the base. Such symbolic carvings highlight the importance of these mushrooms in the cultural practices of the region's indigenous tribes.

SUPERNATURAL POWERS The hoof fungus (*Fomes fomentarius*), found in Europe, North America, Asia, and Africa, produces large fruit bodies in the shape of a horse's hoof (p.250). The Khanty people of west Siberia burnt it at the entrance to burial chambers; the smoke was thought to ward off any influences that the corpse would have on the living.

OLD WORLD RITUALS The ancient Hindu text *Rig Veda*
mentions "soma": an invigorating drink made from a red plant
with no leaves, blossoms, or roots but with meaty stems. In his
book *Soma of the Aryans*, the ethnomycologist Robert Gordon
Wasson suggests that this "plant" was in fact the fly agaric. The
Rig Veda depicts Indra, the king of the gods, and the fire god Agni
with this precious elixir, which was thought to give immortality
and strength to those who consumed it.

Another revered god in the Hindu religion is Shiva, the protector,
who is often portrayed holding a mushroom. Elsewhere in south
Asia, female fertility figurines from the Harappa civilization in the
Indus Valley depict the fly agaric mushroom in their headdresses,
conveying the importance of this mushroom as a spiritual symbol.

RITUALS OF THE SIBERIAN SHAMANS In Siberian
folklore, the fly agaric was celebrated for its hallucinogenic
properties, and it was central to shamanistic rituals. The biologically
active hallucinogen produced in the cap of the mushroom is not
broken down by the body and is therefore excreted in urine.
Folklore recounts that, after ingesting the mushroom, the shaman
would urinate into a vessel and then share it with the rest of the
tribe. The mushroom was also an important commodity (see right).

MUSHROOMS IN THE NEW WORLD Many objects of
pre-Columbian North American art, such as ceramic flasks
and masks, depict the fly agaric in some way, suggesting a
deep-rooted association between the early civilizations of the
New World and this hallucinogenic mushroom. It has even been
suggested that the movement of early peoples from northern
Asia across to the Americas – at a time when ice shelves connected
the two continents – brought established cultures, rituals, and
ceremonies involving this mushroom to the New World.

It's all connected
To the indigenous peoples of northern Siberia, the commodities of birch wood, fly agaric, and reindeer were important and interlinked: birch was used for building and fires, fly agaric is mycorrhizal with birch and appears close to it, and reindeer eat birch. People often carried dried mushrooms in birch-wood boxes, wrapped in reindeer hide.

*Fungi have been used in spiritual practices, as foods,
and as medicines for millennia. But our ancestors also found
practical uses for puffballs – in construction, as tinder
to light fires and even to "smoke out" bees.*

THE VALUE OF ANCIENT PUFFBALLS

USE OF PUFFBALLS BY THE ROMANS

Mature specimens of both *Bovista nigrescens* and *Calvatia utriformis* were identified in deposits from the pre-Hadrianic site in Vindolanda in Northumberland, UK, dating back over 1,900 years. It is likely that these puffballs were collected for their hemostatic properties and were used to stem bleeding, although immature puffballs may have been eaten or used as ping pong balls in Roman games.

INDIGENOUS PEOPLES IN NORTH AMERICA The fruit bodies of basidiomycete puffballs are often spherical, and the spongy white centres brown on maturity, becoming powdery and eventually erupting to release thousands of fungal spores.

To the Blackfoot tribes of North America, puffballs were ethereal. They called them *ka-ka-taos* or "fallen stars". These valued fungi were dried and used as tinder to light fires or burned as incense to ward off evil or to anesthetize bees to enable honey collection. Puffball emblems were even painted onto the canvas coverings of tipis, and puffball circles were painted at the tipi's base, presumably in honour of their value as commodities of fire.

Like the leathery tissue of the hoof fungus (*Fomes fomentarius*, p.250), puffballs could smoulder on a campfire, offering hours of fire protection by scaring away wild animals. To people living on the prairie, where very little wood was available, these fungi were invaluable.

NEOLITHIC COMMUNITIES Evidence from archaeological excavations at the neolithic settlement at Skara Brae on the Orkney Islands, UK, also indicates use of mature puffballs by the communities who once lived there. The brown puffball (*Bovista nigrescens*) was found in midden (waste) material, alongside bones, shells, grain, and clay, which was often packed between stones as insulation, presumably to offer some protection against coastal storms.

Like the birch polypore (*Fomitopsis betulina*, p.250), puffballs were also used as a styptic to stem blood flow, while young specimens were probably eaten. Fire can be carried in a smouldering puffball, so their significance in the Neolithic context is likely similar to that of the Blackfoot tribal practices in North America.

Enigmatic carved stone balls excavated at Skara Brae, possibly representing puffballs, suggests that these communities were devoted to these fungi – similar to the ancient mushroom stones of the Mayans in Mesoamerica.

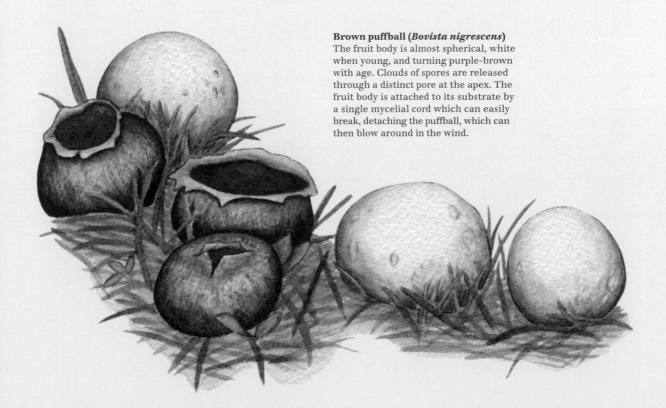

Brown puffball (*Bovista nigrescens*)
The fruit body is almost spherical, white when young, and turning purple-brown with age. Clouds of spores are released through a distinct pore at the apex. The fruit body is attached to its substrate by a single mycelial cord which can easily break, detaching the puffball, which can then blow around in the wind.

Hoof fungus (*Fomes fomentarius*)

HEAL

ANCIENT CULTURES HAVE USED FUNGI for healing for thousands of years. Today, fungi are behind many mainstream medicines – from penicillin and other common antibiotics to medicines that can help combat high cholesterol – and they are being studied as potential forms of treatment for some cancers and mental health conditions.

Turkey tail (*Trametes versicolor*)
This bracket fungus – up to 8 cm (3 in) in diameter –
is common all around the world. It's easy to see where
it gets its common English and American name from:
the fruit body is fan-shaped and resembles a turkey's
tail. It is used in traditional Chinese medicine to treat
infection and inflammation.

Alongside their spiritual significance among traditional communities, mushrooms have long been used for their perceived ability to heal both mind and body.

ANCIENT HEALING POWER OF MUSHROOMS

CHAGA A fungus that grows on birch, chaga (*Inonotus obliquus*) is sometimes referred to as the "gift from god" or the "mushroom of immortality". It has been revered throughout Russia, Siberia, Korea, and China for thousands of years. Found in boreal forests across the northern hemisphere, the fungus grows in living trees, producing a mass of mycelium that erupts through the bark, and has a characteristic burnt charcoal appearance. This is not the fruit body of the fungus but rather a fungal structure called a sclerotium (p.289). It contains chemicals that are believed to have many health benefits, including anticancer, anti-inflammatory, antioxidant, and antidiabetic properties. Powdered extracts of the sclerotium have been used to make teas, soaps, lotions, and essences.

GANODERMA LUCIDUM Known as *lingzhi* in China and *reishi* in Japan, *Ganoderma lucidum* has been used in traditional Chinese and Japanese medicine for at least 4,000 years. In traditional Chinese medicine, *lingzhi* is used to make essences that are believed to promote well-being, particularly to treat liver, kidney, and lung conditions, insomnia, gastric ulcers, and nerve pain. Extracts mixed with skin creams and lotions are said to protect against the sun's rays, and the oil extracted from the fruit body is applied to warts and spots supposedly to act as an antihistamine, soothing the skin.

SHIITAKE MUSHROOMS Growing on the trunks of dead or dying trees such as chestnut or oak, shiitake mushrooms (*Lentinula edodes* p.220) are now well known for their culinary uses. The edible fungus is grown around the world and is commonly available in supermarkets.

But shiitake mushrooms have been cultivated for their perceived healing properties, and used in traditional Chinese and Japanese medicine for centuries. The fungus contains a water-soluble sugar molecule called lentinan, thought to boost immunity. In traditional Chinese medicine, the dried, powdered mushrooms are used to treat arthritis, high cholesterol, diabetes, and to increase stamina.

BIRCH POLYPORE The fruit body of the birch polypore (*Fomitopsis betulina*, p.269) contains antimicrobial chemicals and is thought to have antitumour and anti-inflammatory properties. It was found in the Neolithic travelling kit of Ötzi the Iceman: the 5,000-year-old Neolithic traveller, whose mummified body was discovered in ice in the Ötztal Alps in 1991.

People like Ötzi would probably have held the bracket fungus against a wound to stem blood flow. Combined with other medicinal properties, the fungus would have been a valuable component of the European Neolithic medical kit: thin strips of the fungus were likely used as "Neolithic plasters" to heal scratches and more serious wounds. The fungal fruit body was also used traditionally to make invigorating teas and tinctures with perceived anti-inflammatory properties.

Hoof fungus (*Fomes fomentarius*)
This fungus grows on birch, beech, and other broadleaf tree trunks, producing fruit bodies that resemble horses' hooves. Its spongy tissue can be used to light fires (hence its other common name, tinder fungus) and can also be used to make leather alternatives (p.271).

Caterpillar fungus (*Ophiocordyceps sinensis*)
An ascomycete fungus, this infects ghost moth larvae in the summer while they are feeding on roots of grasses and small plants. The fungus slowly consumes the larvae and doesn't emerge until the following spring. It is known in Tibet as "winter worm summer grass".

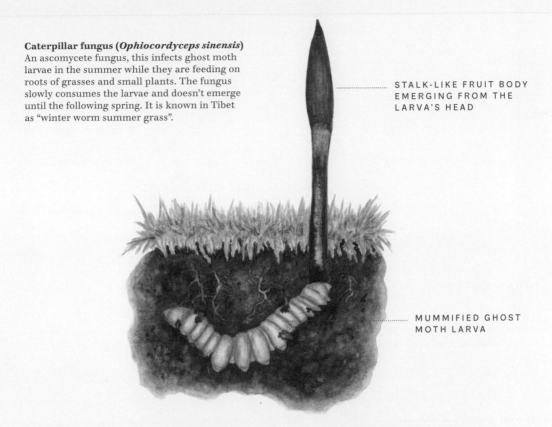

STALK-LIKE FRUIT BODY EMERGING FROM THE LARVA'S HEAD

MUMMIFIED GHOST MOTH LARVA

CATERPILLAR FUNGUS *Ophiocordyceps* are a group of fungi that infect insects and other fungi. The caterpillar fungus (*Ophiocordyceps sinensis*) infects ghost moth larvae, which it kills and mummifies. It then produces a dark, stalk-like fruit body that erupts from the mummified larva's head. Used as a tonic in traditional Chinese medicine, dried extracts of the caterpillar fungus are believed to restore yin and yang. In ancient China, its use was limited to the emperor and members of the emperor's palace. Today, the fungus is highly prized and, as such, is over-picked, threatening its existence (p.113).

HOOF FUNGUS The hoof fungus (*Fomes fomentarius*) is probably one of the earliest medicinal fungi known to be used by humankind. Traditionally used as tinder, traces of the fungus have been recovered from Mesolithic sites dating back over 8,000 years and it, too, was carried by Ötzi the Iceman. The 5th-century physician Hippocrates described the hoof fungus as having cauterizing properties. Later referred to as the "agaric of the surgeons", strips of the fungus were apparently used as bandages by Austrian farmers in the 18th century, by dentists and surgeons to stem bleeding in northern Europe and, in European, west Siberian, and Indian folk medicine, it was allegedly combined with iodine and used as an absorbent dressing for wounds.

Who would have thought that a contaminating mould on a bacterial growth plate would change the course of medicine? The ascomycete Penicillium rubens *did exactly that.*

THE STORY OF PENICILLIN

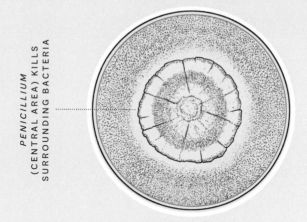

PENICILLIUM (CENTRAL AREA) KILLS SURROUNDING BACTERIA

1928

Alexander Fleming returned from holiday to find that bacterial plates in his laboratory had been contaminated with a fungus, which seemed to have killed the bacteria around it.

1928

Fleming's mycologist colleague Charles La Touche identified the fungus. It became known as *Penicillium rubens*.

1939

Howard Florey and Ernst Chain pioneered a technique to mass-produce penicillin from the fungus.

1945

By 1945, Dorothy Hodgkin had confirmed the chemical structure of the antibiotic.

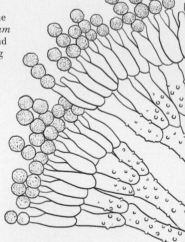

Penicillium close up
When viewed under a microscope, the spore bearing structures of *Penicillium* appear brush-like, often branched and culminating in bulbous ends carrying chains of spores

The discovery of the "wonder drug" penicillin was serendipitous. It was made in 1928 by the bacteriologist Alexander Fleming, who studied the bacterium *Staphylococcus aureus*, which can cause a range of ailments – from minor skin infections to life-threatening conditions such as pneumonia. On his return from holiday, having left a few bacterial plates on his bench, Fleming found that the plates had become contaminated with a fungus. His frustration soon turned to intrigue when he realized that the contaminating fungus appeared to have killed the bacteria surrounding it.

AN ACCIDENTAL MIRACLE Fleming knew very little about fungi, so he asked his mycologist colleague Charles La Touche to look at the plates. La Touche's lab, which was just down the staircase, specialized in culturing fungi from dust samples.

The spores from La Touche's lab were most likely the ones to contaminate the bacterial plates in Fleming's lab. Had the two labs not been in close proximity, and had Fleming not left his bacterial plates out, then the discovery of the most important antibiotic in the history of medicine might never have occurred.

CHANGING THE COURSE OF MEDICINE Penicillins are particularly useful against the bacterium *Staphylococcus* and other bacteria that have high levels of peptidoglycan, a polymer made of sugars and amino acids, in their cell walls. The antibiotic binds to the enzyme involved in forming cross links in peptidoglycan, stopping the bacterial cell wall forming properly, and the bacteria eventually die. Penicillin quickly became known as a wonder drug, providing cures for otherwise fatal diseases, such as septicaemia. However, the widespread use of these antibiotics has led to antibiotic-resistant bacteria, which pose a real threat to human health.

The body naturally rejects "foreign" bodies, such as bacteria or viruses,
but the same response can also cause it to reject transplanted organs.
Fortunately, cyclosporine – a drug produced by a fungus – can help
improve transplant success rates.

THE STORY OF CYCLOSPORINE

A SOIL FUNGUS AND INSECT KILLER

Tolypocladium inflatum is an ascomycete fungus that is found ubiquitously in soil and leaf litter. In nature, this fungus is a decomposer (pp.38–39), but it can also infect insects, including scarab beetle larvae. To reproduce sexually (pp.18–19), the fungus needs to mate with a compatible partner within the insect host. As a pathogen, *Tolypocladium inflatum* uses the cyclosporine it produces to supress the immune system of its larva host so it can invade host tissue and multiply.

A CHANCE DISCOVERY Scientists are always on the lookout for new or "novel" compounds produced by microorganisms that could be used in medicine. This research often involves screening environmental samples, such as industrial effluents, which include discharges from industry as well as soil and water samples. In the late 1960s, a pharmaceutical company in Switzerland was collecting soil samples to screen for novel antibiotics. The team discovered a fungus in a Norwegian soil sample that could produce many interesting chemicals, including one with immunosuppressant properties. The fungus was named *Tolypocladium inflatum*, and the chemical of interest that it produced was called cyclosporine.

HOW DOES CYCLOSPORINE WORK? Cyclosporine suppresses certain types of white blood cell called T-cells. These are part of the body's defence armoury: they stimulate chemical signals in response to the presence of a foreign body, such as an invading microorganism or someone else's heart or kidney, which triggers a defence response to remove it. Cyclosporine stops these chemical signals from being produced by the immune system, which makes it very useful for treating patients after an organ transplant.

*Before **cyclosporine**, very few centres across the world could perform **complex operations** like organ transplants, and **survival rates** were very low. This all changed **thanks to a fungus**.*

Paclitaxel (Taxol), extracted from the Pacific yew tree, is a very valuable cancer drug. However, the extraction process is time-consuming and requires large quantities of yew bark or needles. But maybe fungi can help?

CANCER-FIGHTING FUNGI

WHY DO THESE FUNGI PRODUCE TAXOL?

Yew trees can produce many side branches and, each time one emerges, cracks in the bark can allow pathogens to enter. Inside the tissues of the yew tree, the resident friendly endophytic fungi, such as *Paraconiothyrium*, act like human immunity cells – recognizing and responding to these potential threats by releasing "packets" of Taxol at the wound site, thereby protecting the yew tree from attack. The endophyte does this to protect its niche, at the same time increasing the longevity of its host.

Screening of extracts from the bark of Pacific yew trees (*Taxus brevifolia*) revealed a compound that was highly toxic against human cancer cells. The compound was given the brand name "Taxol" after the tree from which it was extracted. It took several years to purify it and determine its chemical structure. Taxol works by disrupting the way cells divide. Normally, when a cell divides, filaments called microtubules – which act as a sort of skeleton in a cell, giving the cell structure – break and then reform. However, in the presence of Taxol, microtubule production increases and the extra filaments prevent cell division. If the cells are prevented from dividing, they undergo programmed cell death (called apoptosis), which restricts tumour growth.

FUNGAL ENDOPHYTES TO THE RESCUE The demand for the drug is high, yet the availability of sufficient quantities of yew for industrial production is not sustainable. Yew trees are incredibly slow growing, and the process of extracting the drug takes a long time. Scientists discovered that a fungal endophyte (pp.68–69) living within the tissues of the yew tree also produces Taxol. If extraction from fungi grown on an industrial scale can replace the current methods of obtaining the drug, large quantities could be made at relatively low cost and without chopping trees down – making it more environmentally friendly.

Taxol at the site of a yew bud emerging
Wood-decay fungi (yellow in the cross section below) can enter through lesions where the new bud forms and begins to push out from the bark. Fungal endophytes (blue) residing within the tree respond to chemical signals (purple) produced as the wood-decay fungus infects the tree. The fungal endophytes release packages of Taxol (red), which halts the wood-decay fungus in its tracks.

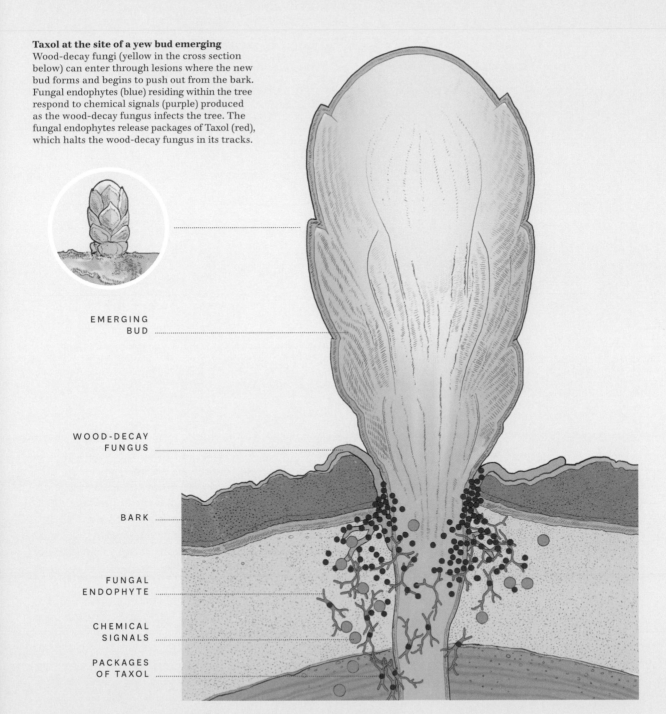

EMERGING BUD

WOOD-DECAY FUNGUS

BARK

FUNGAL ENDOPHYTE

CHEMICAL SIGNALS

PACKAGES OF TAXOL

Mysterious black grains replacing healthy ones,
a deadly, hallucinatory condition – the ergot fungus
has caused chaos for centuries. But science is now
uncovering its medicinal value.

THE ERGOT FUNGUS AND LSD

WHY DOES THE FUNGUS PRODUCE ERGOTS?

Ergots are fungal "survival structures" called sclerotia. Many fungi produce these structures to protect themselves from drying out, UV damage, microbial attack and, in the case of plant pathogens, the absence of a suitable host plant. The sclerotia produced by *Claviceps purpurea* enable the fungus to survive winter in the soil until spring, when young grasses start to emerge and the cycle of infection begins again.

The fungus *Claviceps purpurea* is an ascomycete that infects grasses and replaces healthy grains with strange, blackened structures called ergots. Named after the French "argot" meaning "cock's spur", ergots resemble the spur on the foot of a cockerel. Accidentally consuming ergots causes a debilitating condition called ergotism, which can trigger hallucinations, cause gangrene, and even death. The first mention of ergots appears on an Assyrian tablet around 600 BCE, where they are described as "a noxious pustule in the ear of grain". Peasant farming communities in the Middle Ages were cursed by what they called "St Anthony's Fire" after eating ergot-contaminated rye bread. The condition has even been linked with the Salem Witch Trials of the 1600s: some historians have suggested that those believed to have been bewitched were actually suffering from ergotism. Today, grain is carefully monitored, so outbreaks are rare.

MEDICINAL VALUE Although toxic if consumed, ergots are a treasure trove of valuable compounds known as alkaloids. These are produced by plants, fungi, and other organisms. The exciting thing about ergot alkaloids is that they are structurally similar to the human body's natural chemical messengers (called neurotransmitters), such as serotonin. This means they can be used to mimic chemical messengers in the body. So far, they have been used to treat severe migraine and manage Parkinson's disease. One of the most notable compounds to emerge from the study of ergots is LSD, the controlled use of which looks promising for the treatment of certain mental health conditions.

HOW ERGOTS FORM AND SPREAD The fungus infects young grain florets, producing masses of asexual spores (pp.54–55), which are spread by rain splash and by insects, causing further infections. Eventually, spore production ceases and an ergot full of fungal tissue is formed in place of grain. The blackened fungal structures are harvested with healthy grain or fall to the ground and overwinter until spring.

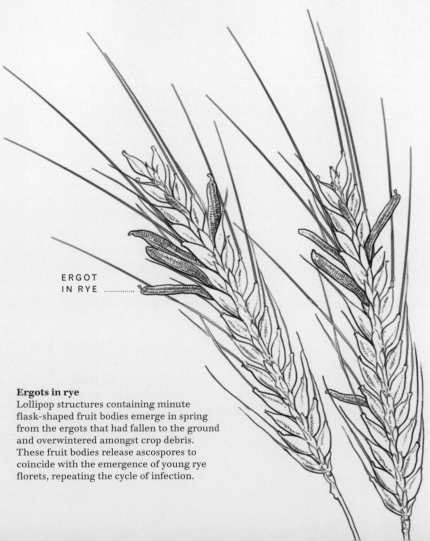

ERGOT
IN RYE ....

Ergots in rye
Lollipop structures containing minute flask-shaped fruit bodies emerge in spring from the ergots that had fallen to the ground and overwintered amongst crop debris. These fruit bodies release ascospores to coincide with the emergence of young rye florets, repeating the cycle of infection.

With its distinctive domed and pointed cap, the liberty cap
(Psilocybe semilanceata) *is normally found in undisturbed*
grassland areas. This common fungus is one of several that
produce the compound psilocybin, a psychoactive compound
that could revolutionize mental health treatment.

PSILOCYBIN AND MENTAL HEALTH

Psilocybin is a secondary metabolite, a compound produced as a defence chemical. A psychoactive compound – affecting the mind – it is thought by some researchers to play a role in certain fungus–insect interactions, altering the behaviour of the insect host so the fungus can spread its spores. Discovered by the same scientist who found LSD (pp.258–259), the hallucinogenic effects of psilocybin in humans, which last for several hours after ingestion, include heightened visual and auditory perception and a feeling of euphoria. It works by binding to the same receptors in the body as the neurotransmitter serotonin and boosting the serotonin response. Serotonin plays an important role in lifting our mood and making us feel happy.

MEDICAL RESEARCH In the 1960s, scientists began to explore the use of psilocybin to treat psychosis and other mental health conditions. However, recreational use of the drug led to the halting of medical trials. Strict laws were put in place surrounding the picking, culturing, and use of *Psilocybe* species, many of which are still in place. Several countries, including the UK, currently list psilocybin as a Class A drug. It is illegal to pick, culture, or consume this fungus.

Fifty years later, there has been a resurgence of interest in the controlled, individualized use of psilocybin to treat anxiety, depression, and other mental health-related disorders. We still have much to understand about the possible therapeutic benefits and dangers of prescribing psilocybin and the potential side effects associated with various dose regimes. Time will reveal its true potential.

**The liberty cap
(*Psilocybe semilanceata*)**
This is one of the fungi that
produces the psychoactive
drug psilocybin.

Brain imagery in response to psilocybin
Functional magnetic resonance images (fMRI) can
be used to measure changes in blood flow between key
areas of the brain in response to psilocybin. The images
show that in the presence of the drug (below left), there
is a decrease in blood flow (red shading) and therefore
a decrease in activity and connectivity between key
regions of the brain compared with a placebo (below
right). It is this decrease in coupling that is thought
to contribute to the uninhibited cognitive effects
experienced by the patient.

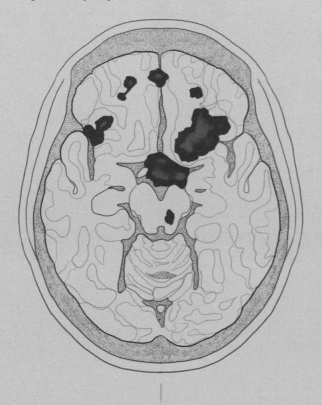

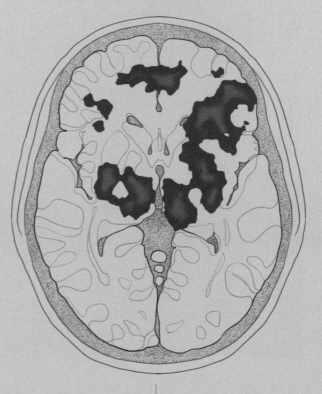

WITH PSILOCYBIN

WITHOUT PSILOCYBIN

Cholesterol is found in the blood and in foods, and is essential for the efficient functioning of cells, but too much can lead to cardiovascular disease. Fortunately, fungi can help keep cholesterol in check.

THE FIGHT AGAINST CHOLESTEROL

Cholesterol is a fatty substance carried around the body in the blood by proteins called lipoproteins. High density lipoproteins (HDLs) are "good" because they remove cholesterol from the body, whereas low density lipoproteins (LDLs) are "bad" as they carry cholesterol around the body. Too much LDL cholesterol can cause a build-up in blood vessels and reduce blood flow. People with high LDL cholesterol are therefore more at risk of cardiovascular disease.

THE SEARCH FOR STATINS In the 1970s, pioneering work inspired by Fleming's discovery of penicillin (pp.252–253) started in Tokyo to screen thousands of fungal cultures for compounds that could inhibit the biochemical pathways that cause the liver to produce cholesterol. These compounds, called statins, are produced by certain fungi to help protect them from competitors. Statins bind to an enzyme usually involved in making cholesterol, stopping the process. The first statin to be identified was compactin (now called mevastatin), produced by the fungus *Penicillium citrinum*. This product never reached the market because the trials used concentrations that were too high. However, in the late 1970s, a second fungal statin called lovastatin, with a similar structure, was identified in *Aspergillus terreus* and, by the late 1980s, it was made commercially available. Many statins now on the market are made synthetically, though some processes still involve fungi. Today, these fungal-derived products vastly improve the lives of over 30 million people who take statins to relieve cardiovascular conditions.

When the search began to find compounds that could **reduce levels** *of blood cholesterol and* **mitigate the risk** *of heart disease, researchers turned* **to fungi**.

Bloodred webcap (*Cortinarius sanguineus*)

CHAPTER X

FUNGI ARE WORKING behind the
scenes in many everyday processes that we take
for granted: fungi dye our clothes, clean our
homes, and give us cheese, wine, and soy sauce.
They might also be the key to a greener future:
fungi can give us sustainable materials and
processes, provide new building materials,
packaging, and paper, and even degrade
unwanted plastics.

Technology can be described as any device, method, or system that applies scientific knowledge to a practical use, and technologies that use fungi – so called mycotechnology – have been around for centuries. But this vast resource is still largely untapped.

MYCOTECHNOLOGY

MYCELIUM SUPERMATERIAL The fine networks of hyphae that form the mycelium – the main body of filamentous fungi – have been dubbed a "supermaterial". Mycelium is very fine but also strong, able to resist water and breaking under pressure, thanks in part to the biopolymer chitin (pp.16–17) in its cell walls. Coordinated groups of hyphae also show great strength – which is why some mushrooms can pop up through tarmac and lift paving slabs. With the right fungus and growing conditions, scientists can manipulate mycelium to grow in a way that replicates a chosen material. For example, it can be used to create sustainable materials for building and packaging (pp.268–269), or to craft lampshades and other furniture. It can also be grown to produce fabrics or as a vegan alternative to leather (pp.270–271), or grown to resemble the structure of meat fibres (pp.272–273).

YEASTS AND ENZYMES Microscopic, single-celled fungi, such as the yeast *Saccharomyces cerevisiae*, have been used in the brewing and baking industries for centuries. As yeast cells are eukaryotic (pp.16–17), they have all the processing "machinery" they need to translate human genetic material into proteins. This makes them suitable for producing pharmaceutical products, such as insulin.

The insulin gene is cloned into a circular piece of DNA called a plasmid. The gene is then expressed in the yeast cell, and the cells produce human insulin. Yeasts can be used to produce hepatitis vaccines and human serum albumin, a protein needed to treat serious blood loss and burn injuries.

Fungi produce an amazing array of chemicals and enzymes to help them break down organic matter. These are used in industry, often providing more environmentally friendly alternatives to traditional methods (pp.276–277). Certain fungi even produce enzymes that have the potential to break down single-use plastics, leaving no microplastic residue – an attribute that has the potential to provide much-needed support in tackling humankind's waste problem.

What's more, the use of certain fungal enzymes in myco-remediation, to help break down oil spills and soak up uranium and other toxic radioactive chemicals, offers great potential in the quest to re-green the planet (pp.280–281).

Mycelium
This shows mycelium growing from a wood block 2 x 2 cm (¾ x ¾ in) across. The individual filaments of this particular fungus have aggregated to form cords (pp.36–37, 118–119) with good tensile strength. Mycelium has many characteristics that make it valuable in mycotechnology, from meat alternatives to biodegradable fabrics.

Fungi will grow on just about anything, including waste.
Mycelium can be grown at scale without chemicals
or high temperatures into any shape or form – from
blocks for the building industry to bespoke packaging.

MYCELIUM IN A SUSTAINABLE WORLD

Given the eco-friendly production and biodegradable nature of mycelium-based materials, they make the perfect choice for industries that are working towards achieving circular economies for their businesses. Alongside its green credentials, mycelium's benefits include rigidity, strength, and fire-retardant properties, thanks to the chitin (p.16) in fungal cell walls. Fungal cells' water-hating proteins called hydrophobins also give mycelium waterproofing qualities.

THE FUTURE IS FUNGAL Designers can use fungal-generated composites to make chairs, lampshades, and other artisan products, as well as clothing (pp.270–271), all of which are fully biodegradable, so it is hardly surprising that mycelium has been hailed as the super-material of the future. The properties of each product depend on the production process, starting material (the material the fungus is grown on), and the fungal species used. The hoof fungus (*Fomes fomentarius*), for example, is grown on green wood waste to make items such as chairs.

Mycelium-derived packaging also provides an alternative to plastic packaging, as well as a carbon-neutral building material to replace existing composites. The edible grey oyster mushroom (*Pleurotus ostreatus*), the bracket hoof fungus (*Fomes fomentarius*), and the medicinal mushroom *reishi* or *lingzhi* (*Ganoderma lucidum*) have already been successfully grown on grain and waste materials to create fungal composite materials with similar strength, thermal, and electrical insulation properties as conventional building materials.

MAKING FUNGAL PAPER

BIRCH POLYPORE
(*FOMITOPSIS BETULINA*)

1.
Many tree-dwelling bracket fungi, such as the birch polypore (*Fomitopsis betulina*), have fibrous fruit bodies, and these fungi can be used to make paper.

2.
The fungal fruit body is soaked in water, chopped into small pieces, and then macerated using a blender, forming a spongy marshmallow mixture.

3.
The mixture is poured onto the surface of a paper-making frame over a tray to catch excess water. A sheet of blotting paper is placed over the pulp and the sheet inverted and frame removed.

4.
The newly exposed side is covered with blotting paper and then the whole fungal paper sheet is compressed.

5.
Once most of the water has been absorbed by the blotting paper, the fungal paper can be air dried or dried with a hairdryer.

6.
Fungal paper resembles paper made from wood, but is made from chitin rather than cellulose.

7.
Different fungi produce different shades of fungal paper, depending on which pigments they produce.

**Bloodred webcap
(*Cortinarius sanguineus*)**
This is a very distinctive small mushroom with a dark red cap, initially convex but flattening with age. The slender, blood-red stem has a fibrous veil (cortina), characteristic of this genus. Often found growing with conifers with which it is mycorrhizal (pp.60–61), the mushrooms of this fungus can be used to produce pink, red, and purple dyes.

FUNGAL DYES

Lichens (pp.70–71) and other fungi produce many different natural pigments that have been used to dye natural fabrics, such as wool and silk, for centuries. The dyer's polypore (*Phaeolus schweinitzii*, p.195) produces a range of colours, whereas the cinnamon bracket (*Hapilopilus nidulans*) produces a purple dye; sulphur tuft (*Hypholoma fasciculare*, p.127), produces a vibrant yellow dye; and the scaly hedgehog fungus (*Sarcodon imbricatus*) produces blue-green dyes.

The vibrant colours, unusual textures, and various shapes and sizes of mushrooms, lichens, and other fruit bodies have inspired humankind for millennia. But more recently, fungi have made it onto the catwalk.

MYCELIUM FASHION

Mycelium can provide sustainable materials such as leather alternatives and natural dyes, prompting fashion designers around the world to turn to the fascinating kingdom of fungi. Stella McCartney's "Mushrooms are the future" collection, for example, features luxury bags made from mycelium leather.

MYCELIUM MATERIALS To craft dresses and other fashion items, mycelium is grown on a template of certain textiles and waste materials, forming a mat of hyphae that is the perfect size and fit. The result is a seamless garment with no need for cutting or sewing – so there are no offcuts and therefore no fabric waste. The fungal fabric can be naturally dyed, has inbuilt flexibility, and is completely home compostable.

For centuries, the spongy interior of the bracket hoof fungus (*Fomes fomentarius*, p.250) has been used in Eastern European countries, such as Romania, and in North America to make a velvety, leather-like material, referred to as amadou, which can be crafted into hats, bags, belts, and other items of clothing. Today, the fungus is grown on recycled sawdust in large temperature-controlled, walk-in incubators, and its fruit bodies are harvested and then processed. Amadou is naturally antimicrobial, a good insulator, lightweight, and flexible, making it an ideal biodegradable, vegan alternative to leather.

Mushrooms are made up of tightly packed fungal filaments called hyphae, which give these fruit bodies their characteristic meaty texture. The same effect can be achieved at scale to produce the meat alternative mycoprotein.

MEAT ALTERNATIVES FROM MYCELIUM

For centuries, mushrooms have been used to give texture to dishes that lack meat. However, it was not until the 1960s – when there were real concerns over whether there was sufficient animal protein to match world demand, particularly in developing countries – that alternative sources of protein were sought. Scientists turned to fungi for the answer. In particular, they wanted to replicate the way hyphae aligned in mushrooms to give them their meaty texture.

The search began to find a filamentous fungus that was safe to eat and could be grown in continuous culture in a fermenter to produce meat-like mycelium. Over 3,000 fungi were screened from soil samples collected worldwide. The most promising was isolated in a field close to High Wycombe in the United Kingdom: the ascomycete *Fusarium venenatum*.

HEALTH BENEFITS Mycoprotein has all the essential amino acids but little fat and no cholesterol, making it a low-calorie alternative to meat. As part of a healthy lifestyle, mycoprotein is thought to help to reduce bad cholesterol. Substituting one meat meal for a mycoprotein alternative each week can also reduce greenhouse gas emissions, which helps to protect the environment.

HOW MYCOPROTEIN IS GROWN

1.
Fusarium venenatum is grown under conditions that prevent excessive hyphal branching but produce uniform hyphal filaments.

2.
This is carried out in large air-lift fermenters that bubble air through the chamber and avoid the use of mechanical mixers, which would damage the hyphae.

3.
As it grows, the mycelium is continuously harvested using a moving bed filtration technique that aligns the hyphal filaments in a way that most closely resembles the texture of meat.

4.
To begin with, the harvested mycelium resembles uncooked pastry.

5.
Once processed to remove RNA (a nucleic acid), egg or agar is added to bind the hyphae together, giving mycoprotein the desired meaty texture. At this point flavourings can be added.

6.
The flavoured paste is heated to make it set. It is then cooled and cut into shapes or minced, depending on the product required.

7.
The product is then frozen and ice crystals accumulate, forcing the hyphae together, which enhances the meat-like texture.

8.
The mycelium is continuously harvested for a month or so until the hyphae branch too much. Then, a new batch is started with fresh inoculum.

You likely have some idea that yeasts are involved in bread-making and in beer and wine fermentation, and perhaps you've heard that fungal filaments are used to make meat alternatives. But fungi play a much bigger role in food manufacturing than first meets the eye.

FUNGI IN FOOD MANUFACTURING

There is much more to fungi and food than edible mushrooms such as shiitake, oyster, and the cultivated button mushroom (pp.222–223). Fungi are involved in ripening soft cheeses and developing the flavour, texture, and colour of blue-veined cheeses. They also help in the fermentation process that leads to the production of soy sauce, tempeh, kefir, kombucha, miso, and even coffee beans. Fungi are responsible for the production of citric acid used to stabilize fizzy drinks and, what's more, they play an important role in the production of chocolate.

FUNGI AND CHOCOLATE The tropical plant *Theobroma cacao* produces clusters of delicate pink flowers on its branches all year round. The flowers are pollinated by insects, enabling pod fruits to develop. Some trees can produce over 50 pods per year, each pod containing enough seeds for a 100 g (3½ oz) bar of chocolate. Certain fungi partner with the cacao tree, helping its growth. Cacao plants would produce fewer cacao pods if mycorrhizal fungi were not present. In addition, microscopic yeasts aid in cacao seed fermentation, giving chocolate its distinctive flavour.

THE ROLE OF FUNGI IN CHOCOLATE PRODUCTION

1.

Mycorrhizal fungi (pp.60–61), including species of *Acaulospora*, *Gigaspora*, and *Glomus*, form beneficial partnerships with the cacao tree, supplying water and nutrients in exchange for plant sugars that are made during photosynthesis.

2.

Fungi also play a role in the flavour profile of chocolate. Several yeasts, including *Saccharomyces cerevisiae*, *Issatchenkia orientalis*, and *Kluyveromyces marxianus*, are vital in starting the process of fermenting the sweet pulp that surrounds the bitter cacao seed.

3.

The yeasts break down sugars, including pectin, within the pulp, creating a juice that drains away to leave air gaps between the seeds, and allowing oxygen to enter.

4.

The yeast provides the perfect environment for aerobic bacteria to work. The bacteria break down ethanol – a by-product of yeast fermentation.

YEAST FUNGI FERMENT
THE PULP AROUND THE SEEDS

5.

Yeasts contribute different flavour compounds that help give chocolate its taste. These are often specific to the region of production.

6.

Without the partnerships with mycorrhizal fungi and microscopic yeasts, chocolate as we know it wouldn't exist.

Fungi produce an array of enzymes to break down dead leaves and other organic matter for food. These enzymes can be produced at scale and have many clever – and surprising – uses, in everything from fashion to detergent.

USING FUNGI IN THE HOME

PRODUCING STONE-WASHED DENIM

Stone-washed denim became fashionable in the 1980s and involved machine washing denim with pumice stones. The abrasive stones removed some of the dye particles from the fabric. However, the procedure was not uniform, and stone abrasion often damaged both the machinery and fabric. There was also the challenge of removing the used pumice stones after the process. Today, a stone-washed appearance is achieved by breaking down cellulose fibres in the denim by using an enzyme that releases dye pigments. The fungus *Trichoderma reesei* is responsible for producing the cellulases, and the process is much better for the environment.

ECO-FRIENDLY ENZYMES Fungi can grow on cheap materials, including plant and animal waste, and produce large amounts of useful enzymes. These are biological catalysts that speed up the rate of reactions. Several fungal enzymes are commercially available for use in industries such as food and dairy, textiles, pulp and paper, detergent manufacturing, pharmaceuticals, animal feed, and biofuel production. Fungal enzymes are better for the environment than chemical catalysts because their reactions do not involve corrosive reactants, high temperatures, or high pressures. The by-products of the reactions are also usually non-toxic.

CELLULASES AND AMYLASES A group of fungal enzymes called cellulases is responsible for breaking down cellulose, a major component of plant cell walls. A major industrial application of these fungal enzymes is stone-washed jeans (see left).

About 90 per cent of all household detergents contain fungal amylases, enzymes that break down starch into sugars. Fungal lipases (which break down fats) and fungal proteases (which break down proteins) are also widely used in household products. These enzymes are produced by fungi, including *Aspergillus* and *Rhizopus*.

HOW FUNGAL ENZYMES ARE ISOLATED FOR USE IN WASHING DETERGENTS

CULTURE MEDIUM WITH DIFFERENT FUNGI FROM AN ENVIRONMENTAL SAMPLE

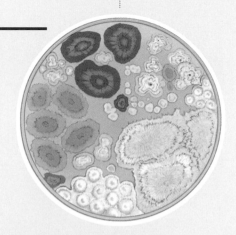

1.

Samples from the environment are screened for fungi that produce high levels of enzymes of value to the detergent industry. These include proteases (to target protein stains), lipases (for fatty stains), and amylases (for starch stains).

2.

Fungi that show the most promise for industrial use are cultured (grown) as individual colonies on agar medium. The most promising strains are then grown individually in a shake flask and levels of enzyme measured.

3.

Fungal cultures that are found to produce high levels of enzymes are used in small-scale bioreactor culture to test how well they perform at a larger scale. Researchers then carry out a test run to prepare for enzyme production.

4.

Fungal cultures are grown in a large-scale fermenter under optimal growth conditions to produce greater quantities of the enzyme.

5.

The culture medium is then harvested and the enzymes are purified ready for use in industrial applications.

Pests can cause devastation to crops, but the ways we try to control pests can be damaging too. Fortunately, certain fungi have proved effective at eradicating some pests – and the side-effects are far less harmful to the environment.

FUNGI FOR PEST CONTROL

Chemicals have been used to control crop pests, from insects to weeds, for many years. They might be good at eradicating the pests, but chemicals can be harmful to the environment, human health, and organisms that get caught in the crossfire. Crop pests can also develop resistance to chemicals, making it even harder to eliminate them. An alternative is biocontrol, which uses microorganisms – such as fungi – rather than chemicals to manage pests.

INSECT PESTS Locusts and other insect pests cause severe crop losses worldwide, and the use of entomopathogenic (insect-killing) fungi (pp.88–89) to kill them is a safe and more environmentally friendly alternative to toxic chemicals as they do not affect non-target organisms. Insect killers such as white muscardine (*Beauveria bassiana*), *Lecanicillium muscarium*, and green muscardine (*Metarhizium anisopliae*) can be mass-produced and have been used as commercial biocontrol agents. Spores are mixed with a powder "carrier" and other components before being applied, often as a spray.

WEED CONTROL Fungi that are pathogens of plants can be used to kill weeds – as long as the chosen fungus kills only the specific weed concerned. The wiry-stemmed rush skeletonweed, for example, is a threat to irrigated farmlands, wheat areas, and rangelands in North

AN INSIDE JOB

Fungi that live within plant cells (pp.68–69) can act as biocontrol agents against plant pathogenic fungi. For example, the fungus *Colletotrichum gloeosporioides* lives within healthy cacao leaves and reduces frosty pod rot infection caused by the basidiomycete fungus *Moniliophthora roreri* by limiting the growth of the invader. This could be because it competes with the pathogen for space or because it produces compounds that limit the pathogen's growth.

America and Australia. It can be controlled using the rust pathogen *Puccinia chondrillina*. The pathogen's limited host range means it kills the weed without affecting other grasses and plants.

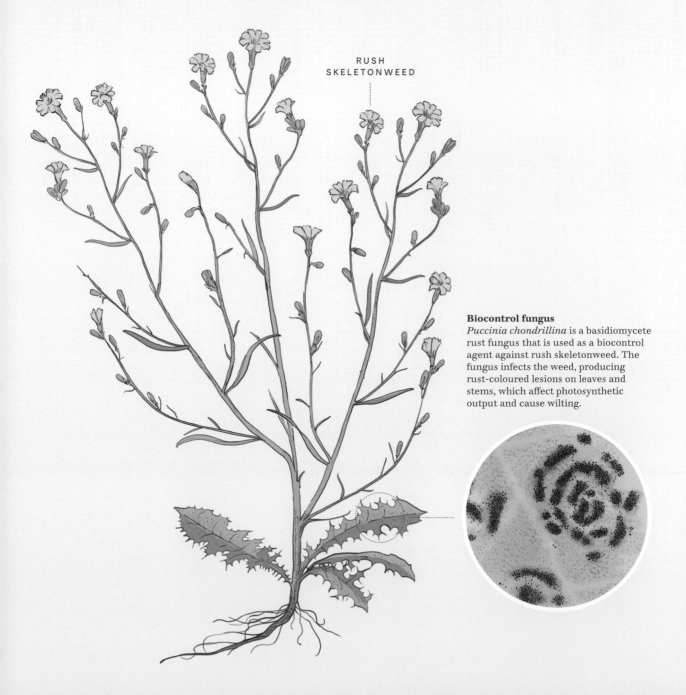

RUSH
SKELETONWEED

Biocontrol fungus
Puccinia chondrillina is a basidiomycete rust fungus that is used as a biocontrol agent against rush skeletonweed. The fungus infects the weed, producing rust-coloured lesions on leaves and stems, which affect photosynthetic output and cause wilting.

Fungi could help solve our waste crisis. In a process called myco-remediation, fungi release enzymes that have the potential to break down just about any form of waste – from oil slicks, radiation, and toxic chemicals to single-use plastics.

RE-GREENING THE PLANET

TACKLING PLASTIC WASTE If our current rate of plastic use continues, there will soon be more plastic by weight in our oceans than fish. Fortunately, certain fungi can help. Research has shown that *Aspergillus tubingensis*, isolated from a dumpsite in Pakistan, can break down polyurethane used to insulate refrigerators. *Penicillium simplicissimum*, found on decaying vegetation, can break down polyethylene, the single-use plastic in carrier bags. Studies are also assessing whether fungi could help break down protective face masks and gloves.

MOPPING UP HARMFUL WASTE The isotope uranium-235 is used in nuclear power plants and modern ammunitions. When it ends up in the soil, for example after a conflict, the contamination can be long-lasting and a risk to human health. But mycorrhizal fungi (pp.60–61) and fungi commonly found in the soil can help to clean up dangerous uranium deposits by making them less active, or by capturing and localizing them within their hyphae.

PAST, PRESENT, AND FUTURE Fungi were instrumental in helping plants to colonize land 450 million years ago. Today, over 90 per cent of plants depend on close associations with fungal partners to provide them with water, nutrients, and other benefits. Decomposer fungi were responsible for the first terrestrial soils, and remain our planet's best recyclers of dead organic matter, releasing valuable nutrients for plants and other organisms to use. Without fungi our planet would not function, and neither would life as we know it.

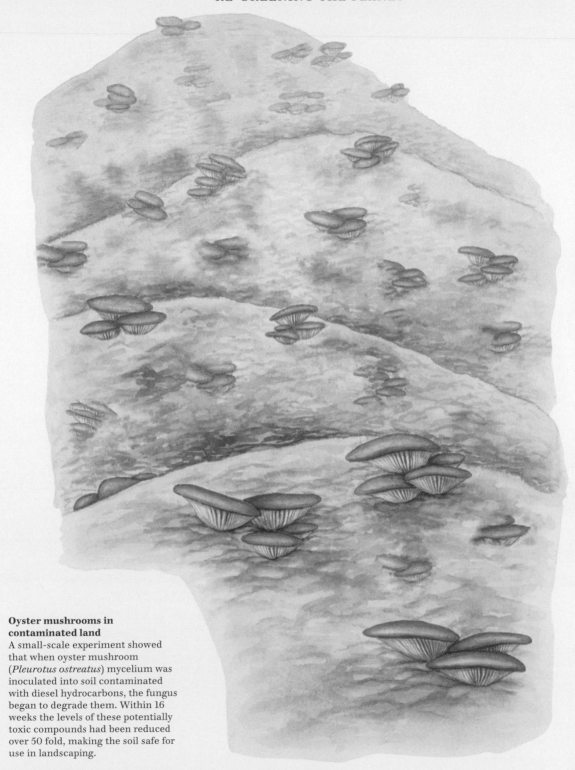

Oyster mushrooms in contaminated land
A small-scale experiment showed that when oyster mushroom (*Pleurotus ostreatus*) mycelium was inoculated into soil contaminated with diesel hydrocarbons, the fungus began to degrade them. Within 16 weeks the levels of these potentially toxic compounds had been reduced over 50 fold, making the soil safe for use in landscaping.

After first recommending general books on fungi, this further reading and resources list is then organized by chapter, with reference to some specific topics covered in each chapter.

FURTHER READING AND RESOURCES

General Introductions

Boddy L & Coleman M (2010) *From Another Kingdom: the Amazing World of Fungi*. Royal Botanic Garden Edinburgh.

O'Reilly P (2022) *Fascinated by Fungi*. Coch-y-Bonddu Books Ltd.

Piepenbring M (2015) *Introduction to Mycology in the Tropics*. The American Phytopathological Association, USA.
A well-illustrated, straightforward introduction to fungal biology. Many examples are of tropical fungi, but the concepts are relevant to all areas.

Pouliot A (2018) *The Allure of Fungi*. CSIRO Publishing.

Roberts P & Evans S (2011) *The Book of Fungi*. Ivy Press.

Seifert K (2022) *The Hidden Kingdom of Fungi: Exploring the Microscopic World in Our Forests, Homes, and Bodies*. Greystone Books.

Sheldrake M (2021) *Entangled Life: How Fungi Make Our Worlds, Change Our Minds, and Shape Our Futures*. Bodley Head.

Detailed textbooks

Boddy L (2021) *Fungi and Trees: their Complex Relationships*. The Arboriculture Association, Stroud.
Contains a basic description of fungal biology, with a plethora of examples of fungi associating with living trees and dead wood.

Kendrick B (2017) *The Fifth Kingdom: An Introduction to Mycology*. Hackett Publishing Company, Inc. (4th edition).

Moore D, Robson GD & Trinci APJ (2020) *21st Century Guidebook to Fungi*. Cambridge University Press.

Watkinson SC, Boddy L & Money NP (2015) *The Fungi*. Academic Press.

CHAPTER I: DISCOVER

Nagy LG et al. *Fungal Tree of Life: Macroscopic Diversity of Fungi*. **group.szbk.u-szeged.hu/sysbiol/nagy-laszlo-lab-poster.html**
A downloadable poster. Further information on: The rise of kingdom fungi pp.12–13; Who belongs to kingdom fungi? pp.14–15.

Puginier C, Keller J & Delaux P-M (2022) Plant–microbe interactions that have impacted plant terrestrializations. *Plant Physiology*, 190, 72–74. doi: 10.1093/plphys/kiac258.
Further information on: Ancient plant partners pp.28–29.

Stephenson SL & Stempen H (2000) *Myxomycetes: a Handbook of Slime Moulds*. Timber Press.

Strullu-Derrien C et al. (2018) The origin and evolution of mycorrhizal symbioses: from palaeomycology to phylogenomics. *New Phytologist*, 220, 1012–1030.
Further information on: Ancient fungi pp.26–27; Ancient plant partners pp.28–29.

Watling R (2010) The hidden kingdom. In: Boddy L & Coleman M, pp.24–33.
Further information relevant to many of the spreads in this chapter.

Wu B, Hussain M, Zhang W, Stadler M, Liu X & Xiang M (2019) Current insights into fungal species diversity and perspective on naming the environmental DNA sequences of fungi. *Mycology*, 10(3):127–140. doi: 10.1080/21501203.2019.1614106.
Further information on: What's in a name? pp.22–23; How many fungi are there? pp.24–25.

CHAPTER II: LIVE

Aleklett K & Boddy L (2021) Fungal behaviour: a new frontier in behavioural ecology. *Trends in Ecology and Evolution*. 36(9):787–796. doi: 10.1016/j.tree.2021.05.006.
Further information on: Mycelium senses pp.40–41; Fungal behaviour and memory pp.42–43.

Watkinson SC, Boddy L, & Money NP (2015) *The Fungi*. Academic Press.
Further information relevant to many of the spreads in this chapter.

CHAPTER III: INTERACT

Boddy L (2021) Beneficial relationships between fungi and trees. In: Boddy, pp.66–83.
Further information on: Fungal partnerships with plants pp.60–61; Wood wide web pp.62–63; Hidden fungi: endophytes pp.68–69; Lichens pp.70–71.

Boddy L (2021) Fungi that harm trees. In: Boddy, pp.124–151.
Further information on: Altering our green landscape pp.78–79; Emerging fungal diseases pp.80–81.

Boddy L (2021) Interactions among tree-associated fungi and with other organisms. In: Boddy, pp.124–151.
Further information on: Partnering with invertebrates pp.82–83; Eat or be eaten pp.86–87; Fungal partnerships with birds and mammals pp.84–85; Fungi and bacteria pp.100–101; Fungus wars pp.102–103.

Boddy L (2015) Interactions between fungi and other microbes. In Watkinson et al., pp.337–360.
Further information on: Fungi and bacteria pp.100–101; Fungus wars pp.102–103.

Boddy L (2015) Interactions with humans and other animals. In Watkinson et al., pp.293–336.
Further information on: Partnering with invertebrates pp.82–83; Eat or be eaten pp.86–87; Mummies and zombies pp.88–89; Amphibian and mammal killers pp.90–91; Fungal allergens pp.94–95; Fungal toxins pp.96–97; Human diseases caused by fungi pp.98–99.

Boddy L (2015) Pathogens of autotrophs. In Watkinson et al., pp.245–292.
Further information on: Fungus diseases and crops pp.72–73; Saving the banana pp.74–75; The gardener's nightmare pp.76–77; Altering our green landscape pp.78–79; Emerging fungal diseases pp.80–81.

Boddy L (2014) Soils of war. *New Scientist* 224, 42–45.
Further information on: Mummies and zombies pp.88–89; Fungus wars pp.102–103.

Boddy L, Dyer P & Helfer S (2010) Plant pests and perfect partners. In: Boddy L & Coleman M, pp.51–65.
Further information on: Fungal partnerships with plants pp.60–61; Wood wide web pp.62–63; Plant cheaters pp.64–65; Lichens pp.70–71; Hidden fungi: endophytes pp.68–69; Fungus diseases and crops pp.72–73.

Cui L, Morris A, & Ghedin E (2013) The human mycobiome in health and disease. *Genome Medicine* 5, 63. doi: 10.1186/gm467
Further information on: The human mycobiome pp.92–93.

Combes M, Weber JF, & Boddy L (2022) So what is ash dieback? *Small Woods*, Summer, 14–16.
Further information on: Emerging fungal diseases pp.80–81.

Deveau A et al. (2018). Bacterial–fungal interactions: ecology, mechanisms and challenges. *FEMS Microbiology Reviews*, 42(3), 335–352.
Further information on: Fungi and bacteria pp.100–101.

Elliott TF, Jusino MA, & Vernes K (2020) Ornithomycology: an overlooked field of study. The ecological significance of symbiotic associations between birds and fungi. bou.org.uk/blog-elliott-ornithomycology/
Further information on: Fungal partnerships with birds and mammals pp.84–85.

Evans HC & Boddy L (2010) Animal slayers, saviours and socialists. In: Boddy L & Coleman M, pp.67–81.
Further information on: Partnering with invertebrates pp.82–83; Mummies and zombies pp.88–89.

Fisher MC et al. (2012) Emerging fungal threats to animal, plant and ecosystem health. *Nature* 484, 186–194.
Further information on: Emerging fungal diseases pp.80–81.

Hiscox JA, O'Leary J, & Boddy L (2018) Fungus wars: basidiomycete battles in wood decay. *Studies in Mycology* 89, 117–124.
Further information on: Fungus wars pp.102–103.

Kolmer JA, Ordonez ME, & Groth JV (2009). The Rust Fungi. In: *Encyclopedia of Life Sciences* (ELS). John Wiley & Sons. doi: 10.1002/9780470015902.a0021264
Further information on: Fungus diseases and crops pp.72–73.

Kumar P, Mahato DK, Kamle M, Mohanta TK, & Kang SG (2017). Aflatoxins: A Global Concern for Food Safety, Human Health and Their Management. *Frontiers in Microbiology*, 7. doi: 10.3389/fmicb.2016.02170
Further information on: Fungal toxins pp.96–97.

Lucas JA (2020) *Plant Pathology and Plant Pathogens*. John Wiley & Sons.
Further information on: Fungus diseases and crops pp.72–73; The gardener's nightmare pp.76–77.

Oldridge SG, Pegler DN, & Spooner BM (1989) *Wild Mushroom and Toadstool Poisoning*. Royal Botanic Gardens, Kew.
Further information on: Fungal toxins pp.96–97.

Purvis W (2000) *Lichens*. Smithsonian.
Further information on: Lichens pp.70–71.

Rodriguez RJ, White JF, Arnold AE, & Redman RS (2009) Fungal endophytes: diversity and functional roles. *New Phytologist*, 182, 314–330. doi: 10.1111/j.1469-8137.2009.02773.x

Smith SE & Read DJ (2008) *Mycorrhizal Symbiosis*. Elsevier.
Further information on: Fungal partnerships with plants pp.60–61; Wood wide web pp.62–63; Plant cheaters pp.64–65.

Watkinson SC (2015) Mutualistic symbiosis between fungi and autotrophs. In: Watkinson et al., pp.205–243.
Further information relevant to: Fungal partnerships with plants pp.60–61; Wood wide web pp.62–63; Hidden fungi: endophytes pp.68–69.

Watling R (1995) *Children and Toxic Fungi: The Essential Medical Guide to Fungal Poisoning in Children*. Royal Botanic Garden Edinburgh.
Further information on: Fungal toxins pp.96–97.

CHAPTER IV: CHANGE

The following provide further information relevant to many of the spreads in this chapter.

Boddy L (2021) Environmental change. In: Boddy pp.226–251.

Boddy L (2015) Fungi, ecosystems, and global change. In: Watkinson et al., pp.361–400.

Boddy L et al. (2014) Climate variation effects on fungal distribution and fruiting. *Fungal Ecology* 10, 20–33.

Minter D (2010) Safeguarding the future. In: Boddy L & Coleman M, pp.143–153.

Vellinga EC, Wolfe BE, & Pringle A (2009) Global patterns of ectomycorrhizal introductions. *New Phytologist*, 118, 960–973.

CHAPTER V: WALK

Anderson P (2021) Grasslands and CHEGD Fungi. **cieem.net/grasslands-and-chegd-fungi/**
Further information on: Ancient grasslands pp.140–141.

Bechara TJH (2015) Bioluminescence: A fungal nightlight with an internal timer. *Current Biology*, 25(7), R283–R285.
Further information on: Fungi that glow pp.124–125.

Boddy L (2021) *Fungi and Trees: their Complex Relationships*. The Arboriculture Association.
Further information on: Signs of forest fungi pp.118–119; Signs in rotting wood and leaves pp.120–121; Ancient or managed forests pp.126–127; Ageing trees pp.130–131; Decaying branches pp.132–133; Woodland rings pp.136–137.

Dix NJ & Webster J (1995) Phoenicoid fungi. In: *Fungal Ecology*. Chapman & Hall.
Further information on: Fire-loving fungi pp.146–147.

Fox S et al. (2022) Fire as a driver of fungal diversity – A synthesis of current knowledge, *Mycologia*, 114(2), 215–241. doi: 10.1080/00275514.2021.2024422.
Further information on: Fire-loving fungi pp.146–147.

Green et al. (2011) Extremely low lichen growth rates in Taylor Valley, Dry Valleys, continental Antarctica. *Polar Biology* 35, 535–541.
Further information on: Antarctica and the Arctic pp.156–157.

Griffith GW & Roderick K (2008) Saprotropic basidiomycetes in grasslands. In: Boddy L, Frankland JC, & Van West P (eds) *Ecology of Saprotrophic Basidiomycetes*. Academic Press.
Further information on: Ancient grasslands pp.140-141.

Kuo H-C et al. (2014) Secret lifestyles of *Neurospora crassa*. *Scientific Reports*, 4, 5135. doi: 10.1038/srep05135.
Further information on: Fire-loving fungi pp.146-147.

Lodge DJ & Cantrell S (1995) Fungal communities in wet tropical forests; variation in time and space. *Canadian Journal of Botany*, 73. doi: 10.1139/b95-402.
Further information on: Tropical rainforests pp.128-129.

Newsham et al. (2021) Regional diversity of maritime Antarctic soil fungi and predicted responses of guilds and growth forms to climate change. *Frontiers in Microbiology*, 11, 615659. doi: 10.3389/fmicb.2020.615659.
Further information on: Antarctica and the Arctic pp.156-157.

Putzke J et al. (2011) Agaricales (Basidiomycota) fungi in the South Shetland Islands, Antarctica. In INCT-APA Annual Activity Report Science Highlights Thematic Area 2. doi: 10.4322/apa.2014.065.
Further information on: Antarctica and the Arctic pp.156-157.

Spooner B & Roberts P (2005) Dunes and heathland. In: *Fungi*. Collins, pp.290–307.
Further information on: Sand dunes pp.154-155.

Spooner B & Roberts P (2005) Freshwater. In: *Fungi*. Collins, pp.308–329.
Further information on: Freshwater fungi pp.150-151; Marshy habitats pp.148-149.

Spooner B & Roberts P (2005) Grass and grassland. In: *Fungi*. Collins, pp.213–234.
Further information on: Grassland: fairy rings pp.138-139; Ancient grasslands pp.140-141; Gardens and lawns pp.142-143.

Spooner B & Roberts P (2005) Marine and salt marsh. In: *Fungi*. Collins, pp.330–348.
Further information on: Seas and oceans pp.152-153.

Spooner B & Roberts P (2005) Specialised natural habitats. In: *Fungi*. Collins, pp.349–392.
Further information on: Herbivore dung pp.144-145; Fire-loving fungi pp.146-147; Antarctica and the Arctic pp.156-157; Caves and mines pp.158-159.

Online resources

Fungi in Svalbard (2018). Learning Arctic Biology website. **learningarcticbiology.info/learning-arctic-biology/species-and-adaptations/fungi/fungi-in-svalbard/**
Further information on: Antarctica and the Arctic pp.156-157.

Rainforest. **education.nationalgeographic.org/resource/rain-forest**
Further information on: Tropical rainforests pp.128-129.

Rainforest canopy structure. **rainforests.mongabay.com/0202.htm**
Further information on: Tropical rainforests pp.128-129.

Wax cap grassland fungi – a guide to identification and management. PlantLife **https://www.plantlife.org.uk/wp-content/uploads/2023/03/Waxcaps_GrasslandFungiGuideManagement.pdf**
Further information on: Ancient grasslands pp.140-141.

CHAPTER VI: FIND

The form groups illustrated in this chapter are based on those described by Læssøe & Petersen (2019). Their identification guide is based on fungi in temperate Europe, but the form groups apply globally. However, different fungal species are often found in different regions, so it is essential to use literature appropriate to the region you are in. Joining a local mycological group run by experts is often a good way to begin learning how to identify fungi. National mycological associations, such as the British Mycological Society and the North American Mycological Society, can be useful for information on local groups, identification, and other educational resources; they often run social media groups to help with identifications. They sometimes also offer advice on microscopy and staining, which is often needed for species identification. This list of resources is not comprehensive, and we do not endorse any particular product, nor attest to its safety or accuracy.

Identification guidebooks

Buczacki S (2012) *Collins Fungi Guide*. Collins.

Humphries D, Wright C (2021) *Fungi on Trees: A Photographic Reference*. The Arboriculture Association.

Læssøe T & Petersen JH (2019) *Fungi of Temperate Europe*. Volumes 1 and 2. Princeton University Press.

Phillips R (2006) *Mushrooms*. Pan.

Online resources

British Mycological Society. **britmycolsoc.org.uk**
Educational resources, information on fungal conferences, details of some local fungus groups.

British Mycological Society Facebook group. **facebook. com/groups/18843741618**

Colour chart. **mycokey.com/MycokeyDK/DKkeysPDFs/ DanishMycologicalSocietycolourchart.pdf**
Petersen JH. 1996. The Danish Mycological Society's Colour-chart.
Downloadable PDF of a colour chart used for fungus identification and naming.

Cornell mushroom blog **blog.mycology.cornell.edu/**
A place to seek help with in-depth identifications.

First Nature. **first-nature.com**
Detailed descriptions of a wide range of fungi common in Europe and, to some extent, North America.

Fungi Education. **fungieducation.org/**
An introductory resource covering a range of topics, including educating children about fungi.

Fungi of Great Britain and Ireland. **fungi.myspecies.info/**
Source of literature, projects, and a wide range of information.

Global Biodiversity Information Facility. **gbif.org/ species/5**
Global maps of distribution of many species.

This resource was used in combination with others to provide information on species distribution. However, please note that databases are reliant on input from recorders whose identifications may not always be correct. Sources are also sometimes contradictory.

iNaturalist.org. **inaturalist.org/**
A place for keeping and browsing records, getting help with identifications, and communicating with others.

Index Fungorum. **indexfungorum.org/names/names.asp**
Provides the most up-to-date names of fungi, and synonyms.

iSpot Nature. **ispotnature.org.**
A place for keeping and browsing records, getting help with identifications, and communicating with others.

Mushroomexpert.com. **mushroomexpert.com/**
Contains information for studying and identifying fungi.

MycoBank. **mycobank.org**
An online database for the mycological community.

Mycological Society of America online teaching resources. **msafungi.org/website-and-useful-resources**
Contains a range of educational resources, including video links.

MycoWeb **mykoweb.com/**
Contains a range of educational resources, including descriptions of North American fungi.

North American Mycological Association. **namyco.org**
A broad-ranging resource including educational material and details of associated clubs.

SciStarter Citizen science blog. **blog.scistarter. org/2022/08/the-largest-ever-fungi-bioblitz-is-here/**
Contains information on citizen science projects in North America.

Species Fungorum. **speciesfungorum.org/**
A global database of current species names and their relationships to older names.

The Fungarium Kew. **kew.org/science/collections-and-resources/collections/fungarium**
Contains information on Kew's extensive collection of fungi and how to access it.

Tom Volks mushrooms. **botit.botany.wisc.edu/toms_ fungi/**
Educational resource by the renowned American mycologist and educator, the late Tom Volks.

UK Fungus Day. **ukfungusday.co.uk/**
An annual celebration of fungi; the website offers a pointer to autumn events and attractions.

CHAPTER VII: GROW

The following provide further information relevant to many of the spreads in this chapter.

Hickey P (2010) Growing edible fungi. In: Boddy L & Coleman M (eds) *From Another Kingdom: the Amazing World of Fungi*, pp.121–129. Royal Botanic Garden Edinburgh.

Stamets P (1996) *Growing Gourmet and Medicinal Mushrooms*. Ten Speed Press.

CHAPTER VIII: CELEBRATE

The following provide further information relevant to many of the spreads in this chapter.

Harding P (2008) *Mushroom Miscellany*. Collins.

Kiernan H (2010) Fungal monsters in Science Fiction. In Boddy L & Coleman M, pp.105–119.

Rutter G (2010) Fungi and humanity. In Boddy L & Coleman M, pp.93–103.

Spooner B & Roberts P (2005) Folklore and Traditional use. In: *Fungi*. Collins, pp.454–475.

CHAPTER IX: HEAL

The following provide further information relevant to many of the spreads in this chapter.

Atila F, Owaid MN, & Shariati MA (2021) The nutritional and medical benefits of *Agaricus bisporus*: A review. *Journal of Microbiology, Biotechnology and Food Sciences*, 7, 281–286.

Lowe H et al. (2021) The therapeutic potential of psilocybin. *Molecules*, 26, 2948. doi: 10.3390/molecules26102948

Miller H (2001) The Story of Taxol: Nature and Politics in the Pursuit of an Anti-Cancer Drug. *Nature Medicine*, 7, 148. doi: 10.1038/84570

Rogers R (2011) *The Fungal Pharmacy: The complete guide to medicinal mushrooms and lichens of North America*. North Atlantic Books.

CHAPTER X: USE

The following provide further information relevant to many of the spreads in this chapter.

Agrawal BJ (2017) Bio-stoning of denim: An environmental-friendly approach. *Current Trends in Biomedical Engineering & Biosciences*, 3(3), 45–47.

Hyde KD et al. (2019) The amazing potential of fungi: 50 ways we can exploit fungi industrially. *Fungal Diversity*, 97(1), 1–136.

Money NP (2015) Fungi and biotechnology. In: Watkinson et al., pp.401–424.

Niego et al. (2023) The contribution of fungi to the global economy. *Fungal Diversity*. doi.org/10.1007.s13225-023-00520-9

Schwan RF & Wheals AE (2004) The Microbiology of Cocoa Fermentation and its Role in Chocolate Quality, *Critical Reviews in Food Science and Nutrition*, 44(4), 205–221. doi: 10.1080/10408690490464104

Spooner B & Roberts P (2005) Food and technology. In: *Fungi*. Collins, pp.456–514.

The supermarket challenge: **davidmoore.org.uk/assets/fungi4schools/Documentation/POSTERS/display_posters/Supermarket_challenge02.pdf**
These posters offer information on supermarket products that contain fungi, chemicals produced by them, or are produced with the aid of fungi.

Wainwright M (2010). Amazing chemists. In: Boddy L & Coleman M, pp.83–91.

GLOSSARY

adnate (gills) broadly attached to the stem

adnexed (gills) narrowly attached to the stem

amyloid contains starch, such as in some fungus spores

ascomycete informal name for a fungus in the phylum Ascomycota, e.g., a cup fungus

ascospore sexual spore produced by ascomycete fungi

ascus (pl. asci) from the Greek meaning leather bag; microscopic sacs containing ascospores in ascomycete fruit bodies

basidiomycete informal name for a fungus in the phylum Basidiomycota, e.g. a mushroom or bracket

basidiospore sexual spore produced by basidiomycete fungi

basidium (pl. basidia) the cells in basidiomycete fruit bodies on which the basidiospores are made

biotroph a fungus that obtains some or all of its nutrition by intimately associating with other living organisms; it can be detrimental to the other organism or mutually beneficial

bracket a shelf-like, often tough basidiomycete fruit body found on standing trunks and fallen wood

clamp connection a short, backwardly directed side branch on the outside of a basidiomycete hypha, going from the cell on one side of a septum to the cell on the other side

concentric circles that have the same centre

convex curved, rounded outwards

cords/mycelial cords a linear mass of hyphae forming a visible string-like structure, often seen extending from wood beneath leaves on the forest floor

crozier cell a cell at the base of an ascus, shaped like a simple hook-shaped crook

cystidium (pl. cystidia) a cell found between basidia in basidiomycete fruit bodies, often with a distinctive shape, helpful for identification

decurrent (gills) continuing down the stem

diatom a microscopic alga

ectomycorrhiza the structure that forms in one type of partnership between a fungus and the fine roots of trees

ellipsoidal (spores) resembling an ellipse or oval

endophyte a microbe that lives within plants

emarginate (gills) more or less the same length but abruptly shorter near the stem

eukaryote any organism whose cells have a nucleus, e.g. plant, animal, or fungal cells

exudate a fluid that leaks out of an organism, e.g. from a hypha, root, or leaf

fertile layer (hymenium) part of an ascomycete or basidiomycete fruit body that contains the spore-producing cells

floccose covered with woolly tufts

free (gills) not joined to the stem

fruit body a multi-cellular (as opposed to single-celled) fungus structure that houses, supports, and protects the cells that produce sexual spores, e.g., a mushroom, bracket, or cup

fungus (pl. fungi) not a plant, animal, or bacterium but a member of the eukaryotic kingdom fungi, e.g., yeasts, moulds, mushrooms, and lichens

genus (pl. genera) a taxonomic term used in biology; organisms in the same genus are all very closely related in evolutionary terms. Fungi have two-word scientific names: the first is the name of the genus and the second is the name of the species, e.g. *Agaricus campestris* (field mushroom),

Agaricus bisporus (commercial mushrooms), and *Agaricus arvensis* (horse mushroom)

gills vertical plates (lamellae) that hang below the cap of a mushroom and on which spores are produced

hymenium see "fertile layer"

hypha (pl. hyphae) a fine branching filament, which in most fungi is divided into cells/compartments by septa, one cell wide

inoculum part of a fungus that can initiate colonization of dead matter or living organisms

lichen an organism made up of algae or cyanobacteria cells living among hyphal filaments

macroscopic visible to the naked eye, as opposed to microscopic

metabolite any chemical made when an organism breaks down food

motile able to move independently, e.g., by swimming

mushroom the fleshy reproductive structure (fruit body) of some types of fungi, usually produced above ground on the soil surface or whatever the fungus mycelium is feeding on, e.g., wood or fallen leaves

mycelium (pl. mycelia) a network of hyphae that forms the body of most fungi

mycobiont the fungus partner in a lichen

operculate with an operculum

operculum a cap or lid at the tip of an ascus that opens to release the spores

ornamentation (spores) decorative elements, i.e., warts

paraphyses hair-like hyphal filaments in ascomycete fruit bodies, which are interspersed among the spore-producing asci

pathogen an organism that causes disease

perithecium (pl. perithecia) tiny flask-shaped fruit bodies of some ascomycetes

photobiont the alga or cyanobacteria partner in a lichen

phylum a level of biological classification, e.g., birds

resupinate a flattened fruit body attached to a substrate

rhizomorph a linear mass of hyphae that forms a tough structure that grows from the tip, is usually dark red, brown, or black, and looks a bit like a root

sclerotium a fungal survival structure formed from a mass of hyphae, with a thick protective outer rind of hyphae, which helps the fungus survive in difficult conditions

septum (pl. septa) a cross wall that divides hyphae, usually in ascomycetes and basidiomycetes, into compartments or cells; it is occasionally produced in other fungi

sinuate (gills) notched at the end that attaches to the stem

spore the reproductive cell of fungi, equivalent to seeds in flowering plants. There are two types: asexual spores, which are identical to the parent, can be produced on mycelia that have not mated and sometimes in those that have; sexual spores, which contain a mixture of genes from the two parents, can be produced after mating

sterigmata projections from basidia on which the basidiospores are borne

stipe the stem of a fruit body

stroma (pl. stromata) a tough plate-, ball-, or club-shaped mass of mycelium in which tiny flask-shaped fruit bodies of some ascomycetes are formed

substrate the material on which a fungus grows

taxonomy the science of describing and naming organisms, and grouping related organisms together (classifying them)

toadstool a colloquial name for a mushroom-shaped fruit body

translucent semi-transparent, allowing some light through

truffle underground, roughly ball-shaped fruit body produced by some ascomycetes and a few basidiomycetes

umbo a round protuberance at the centre of a mushroom cap

verrucose (spores) with a warty surface

volva sac-like structure at the base of the stem of some agarics

INDEX OF FUNGI

Burgundydrop bonnet
(*Mycena haematopus*)

INDEX

ABOUT THE AUTHORS

LYNNE BODDY MBE is Professor of Fungal Ecology at Cardiff University. As well as writing more than 300 scientific articles, she has authored three academic books on fungi and *Humongous Fungus* (DK) for children. She has been president of the British Mycological Society, and has won many awards for her scientific research from scientific societies in the UK and elsewhere. Lynne has made multiple radio and television appearances, including the BBC's *Deep Down and Dirty*, *The One Show*, *Winter Watch*, and Channel 4's *Sunday Brunch*.

ALI ASHBY is a fungal biologist with a passion for promoting fungal science and for communicating her fascination for the breadth and complexity of the fungal world to a global audience of all ages. She is a former Royal Society University Research Fellow having been based in the Department of Plant Sciences, at the University of Cambridge, where she researched fungal sexual development and fungal diseases of crops. Ali is a member of the Royal Society of Biology (RSB) and the British Mycological Society (BMS); and as a member of BMS Council, and chair of the BMS Fungal Education and Outreach committee, she was instrumental in founding the BMS's flagship event UK Fungus Day, which runs annually throughout the UK. She is deeply involved in fungal education and has recently written the book *Unravelling the Fungus Among Us*.

Horn of plenty
(*Craterellus cornucopioides*)

ACKNOWLEDGEMENTS

The authors would like to thank the many people who have engaged in discussions with them about fungi throughout the course of their careers. It is through these many conversations – with colleagues and friends, academics, educators, business partners, schools, and the general public – that they came to realise how fascinated and intrigued in the fungal kingdom people are. It was, in part, the wish to indulge this broad curiosity that motivated them to write this book. The authors would also like to thank their respective families for their support, and particularly for their advice on what to include in the book – there were so many fascinating topics that could have been covered, so selecting which ones to include was a huge challenge!

The authors thank the many colleagues who shared imagery that formed the basis of briefs for the illustrators, and with special thanks to Pat O'Reilly and First Nature. The authors would also like to thank Thomas Læssøe and Jens H. Petersen for their informative volumes of *Fungi of Temperate Europe* on which the form groups of fungi within the FIND section of this book are based. They also wish to thank the DK editorial team for their relentless hard work and patience in dealing with their queries. To Alice and Jane for their brilliant editorials, Vicky for her skill with page design, Barbara and Sophie for their understanding and resolve when challenges arose and their professional command of the workflow, and to our illustrators: Stuart, Aman, and Dan for producing some wonderful drawings that bring the book to life.

Finally, they would like to extend thanks to Rich Wright, for checking some of the text on identifying features, and checking up-to-date species names in Species Fungorum, Kevin Newsham for advice on Arctic and Antarctic fungi, Peter Crittenden for advice on lichens, and Paul Kirk for providing up-to-date information on numbers of named species of fungi.

DK ACKNOWLEDGEMENTS

The publisher would like to thank Dawn Titmus for editorial assistance, Katie Hewett for proofreading, Ruth Ellis for the index, and Amy Cox, Louise Brigenshaw, and Marianne Markham for their design work.

PICTURE CREDITS

The publisher would like to thank the following for their kind permission to reproduce their illustrations and photographs:
(Key: a-above; b-below/bottom; c-centre; f-far; l-left; r-right; t-top)
Ali Ashby: 173; **Aman Sagoo:** 1, 2, 4, 6, 8, 10, 20, 22, 25, 34, 38, 40b, 47, 48, 52, 57, 58, 64, 66, 70–71, 76–77, 78–79, 83, 85, 86, 96, 102–103, 104, 107, 108 tr, 108tl, 110–111, 112, 114, 116, 118–119, 121, 124, 127, 128, 130–131, 132, 135, 137, 138, 141, 143, 144, 147, 148, 153, 155, 157, 159, 160, 163, 165, 167, 170, 174tr, 174br, 178, 214, 217, 219, 221, 225, 228, 231, 232, 237, 240, 243, 245, 246, 248, 250, 251, 261t, 264, 269, 270, 281; **Dan Crisp:** 180, 182, 183, 184, 185, 186, 187, 188, 190, 191, 192, 194, 195, 196, 198, 200, 201, 202, 204, 205, 206, 208, 209, 210, 212, 213; **Getty Images:** Corbis Historical / Photo Josse / Leemage 235; **Science Photo Library:** Richard Bizley (adapted from) 29tr; **Stuart Jackson-Carter:** 117, 19, 27, 29, 37, 40t, 40l, 43, 45, 50–51, 55, 63, 68, 69, 72–73, 75, 81, 89, 92, 95, 101, 108b, 122, 151, 171, 174tl, 174bl, 177, 223, 227, 252, 253, 257, 259, 261bl, 261br, 267, 275, 277, 279
All other images © Dorling Kindersley

REFERENCES

43 Dowson, Christopher & Rayner, Alan & Boddy, Lynne. (1986) Outgrowth Patterns of Mycelial Cord-forming Basidiomycetes from and between Woody Resource Units in Soil. *Microbiology*-sgm. 132. 203-211. 10.1099/00221287-132-1-203. Adapted from Fig. 2, p.206.

55 Courtesy of Prof. Dr Meike Piepenbring

108 Vellinga, EC, Wolfe, BE and Pringle, A (2009) Global patterns of ectomycorrhizal introductions. *New Phytologist*. 181: 960-973. https://doi.org/10.1111/j.1469-8137.2008.02728.x. Adapted from Fig. 3 (map).

151 Oliveira Fiuza, Patrcia & Cantillo, Taimy & Gulis, Vladislav & Gusmao, Luis. (2017) Ingoldian fungi of Brazil: Some new records and a review including a checklist and a key. *Phytotaxa*. 306. 171–200. 10.11646/phytotaxa.306.3.1. Adapted from Fig. 4.

177 Kirk, PM, Cannon, PF, Minter, DW & Stalpers, JA (2008) *Dictionary of the Fungi*, 10th edition. Wallingford: CABI Publishing. ISBN-10: 0851998267, ISBN-13: 978-0851998268. Reproduced with permission of the Licensor through PLSclear. Adapted from Fig. 10A p.81, and Fig. 14 p.189.

177 Redrawn from Moore, D, Robson, GD & Trinci, APJ (2020). *21st Century Guidebook to Fungi*, 2nd Edition. See p.576. Cambridge, UK: Cambridge University Press. ISBN: 9781108745680 (Basidium B).

Senior Editor Sophie Blackman
Senior Designer Barbara Zuniga
Senior Production Editor Tony Phipps
Senior Production Controller Samantha Cross
Jacket Designer Barbara Zuniga
Sales and Jackets Coordinator Abi Gain
Editorial Manager Ruth O'Rourke
Art Director Maxine Pedliham
Publishing Director Katie Cowan

Editorial Alice Horne and Jane Simmonds
Design Vicky Read
Illustration Aman Sagoo, Stuart Jackson-Carter, and Dan Crisp
Design styling concept Giulia Garbin

First published in Great Britain in 2023 by Dorling Kindersley Limited
DK, One Embassy Gardens, 8 Viaduct Gardens, London, SW11 7BW

The authorised representative in the EEA is Dorling Kindersley Verlag GmbH. Arnulfstr.
124, 80636 Munich, Germany

A CIP catalogue record for this book is available from the British Library.
ISBN: 978-0-2416-1296-5

Printed and bound in China

Disclaimer

Many fungi are poisonous, with effects ranging from stomach cramps to organ failure and
death. Collection for consumption or study is entirely at the reader's own risk. Fungal spores
can cause allergic reactions and inhalation should be avoided. The information presented in
this book is not intended to diagnose, treat, cure, or prevent disease. The illustrations in this
book are artistic representations, and written descriptions are not comprehensive, and
should not be relied upon for accurate identification. Just because a fungus is not mentioned
as being poisonous does not mean that it is safe to consume. The authors and publishers
disclaim, as far as the law allows, any liability arising directly or indirectly from the use
or misuse of the information contained in this book.

www.dk.com

This book was made with Forest
Stewardship Council™ certified
paper – one small step in DK's
commitment to a sustainable future.
For more information go to
www.dk.com/our-green-pledge

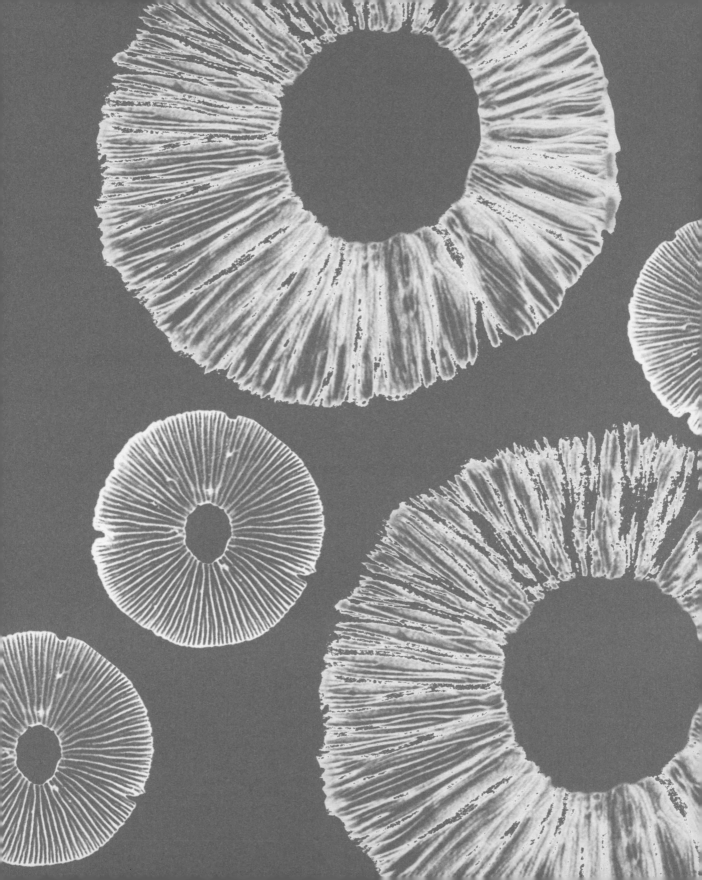